Old Time Radios!

Restoration and Repair

Joseph J. Carr

TAB TAB BOOKS
Blue Ridge Summit, PA

FIRST EDITION
SECOND PRINTING

© 1991 by **TAB Books**.
TAB Books is a division of McGraw-Hill, Inc.

Printed in the United States of America. All rights reserved. The publisher takes no responsibility for the use of any of the materials or methods described in this book, nor for the products thereof.

Library of Congress Cataloging-in-Publication Data

Carr, Joseph J.
    Old time radios! : restoration and repair / by Joseph J. Carr.
        p.   cm.
    Includes index.
    ISBN 0-8306-7342-3   ISBN 0-8306-3342-1 (pbk.)
    1. Radio—Repairing—Amateurs' manuals. I. Title.
TK9956.C3534   1990
621.384′187—dc20   90-44411
                                                        CIP

TAB Books offers software for sale. For information and a catalog, please contact
TAB Software Department, Blue Ridge Summit, PA 17294-0850.

Acquisitions Editor: Roland S. Phelps
Technical Editor: Dan Early
Production: Katherine G. Brown
Book Design: Jaclyn J. Boone

Contents

Acknowledgments

ACKNOWLEDGMENTS FOR PERMISSION TO USE COPYRIGHT MATERIAL GO TO RCA CONSUMER Electronics Division for use of material from *The Receiving Tube Manual,* and the American Radio Relay League (225 Main Street, Newington, CT 06111) for material from the 1952 edition of their incomparable *The Radio Amateur's Handbook.* I also wish to thank John and Ginny Struck, who typed a major portion of the manuscript for me, and Joe Koester of the Mid-Atlantic Antique Radio Club, who allowed me to photograph radios from his collection for illustrations in this book.

Introduction

IT IS A LITTLE DISCONCERTING FOR A GUY AS YOUNG AS ME TO WALK THROUGH A HAMFEST OR antique radio club meeting and see radio receivers *that I worked on when they were under warranty* described as *antiques* or *classics!* But classic and antique radios are very popular today, and many thousands of people now collect and restore these golden oldies. Although I commenced my radio repair career in 1959, I worked on radios from a 1923 model up to the latest models (*latest* as of nearly twenty years ago). The period that I serviced radios spanned the vacuum tube age to the earliest years of the integrated circuit age. When you read about troubleshooting radios of any era in this book, you are getting the story from one who was really there.

After the initial shock of advancing age, I discovered that only a few people still have the old skills needed to keep these sets in good repair and working properly. But once you read this book, you will also be an initiated member of the small (but hopefully growing) band who can make those old hummers work again. It's relatively easy for almost anyone, even though tubes (and now even transistors) are a thing of the past.

The first chapter of this book is a brief history of radio receiver design, starting with the Branly coherers and tappers that Marconi and other early pioneers used, up to the modern superheterodyne receiver. For a while during the 1920s, a number of different designs were offered. Most of these designs are covered in chapter 1. We will also take a quick look at radio receivers over the decades in chapter 2, without going into too much detail. In chapter 3 we discuss the basic theory of vacuum tubes. It is here that we will recover an ancient technology that is needed to understand classic and antique radios.

Chapters 5 through 8 go into detail on vacuum tube circuits found in radio receivers. You will learn about RF amplifier, mixer, local oscillator and converter circuits, and then in chapter 7 we will move on to IF amplifier circuits. In chapter 8 we cover the automatic gain control (AGC) and various detector circuits. Various audio amplifiers and preamplifiers are covered in chapter 9. This discussion will range from the early headphone amplifier to later designs that approached high fidelity performance. Loudspeakers are also covered in the audio chapter. In chapter 5 you will find an extensive discussion of power supply circuits used in radios.

But I'm sure you are interested in diving into the material. Antique and classic radios are a growing and very engaging hobby, and those who can repair or rebuild these sets are especially valued in the community of antique radio fans.

1
A Brief History
of Radio Receivers

RADIO AND TELEVISION BROADCASTING ARE SO WIDESPREAD TODAY, AND HAVE BECOME so familiar, that we often take them for granted. Live satellite coverage from all corners of the earth is routine. But at one time radio—even local radio—was mysterious and exciting. Few things caught the public imagination in the early 1920s like radio broadcasting. From a few enthusiasts—often as not seen as cranks by their friends and neighbors—radio grew to an immense industry in only a few thrilling decades.

Early radio was called *wireless* (and still is in some parts of the world) because it allowed low-distance communications like the telegraph, but without the vulnerable wires strung from point to point. Wireless telegraphy also allowed service to points not normally open to landline telegraphy, such as remote mountain areas and ships at sea. In fact, it was probably mariners who first showed real interest in the practical use of wireless telegraphy.

Although wireless research and development started in the nineteenth century, the dream of wireless telegraphy and telephony eluded experimenters for many years. However, advances were made, albeit slowly, so that by the turn of the twentieth century the technology was ready for a period of fast-paced, almost startling progress. Oddly enough, much of the early radio industry was "garage-born," and indeed some radio companies manufactured equipment in such informal quarters up until the 1930s.

Nineteenth-Century Experiments

One of the earliest, possibly the very first, example of wireless communication occurred in the United States in 1865. On a mountain in West Virginia, a Washington, D.C. dentist named Mahlon Loomis flew a kite that carried a large square of copper "gauze." A wire connected the kite to a grounded galvanometer below. On another mountain 18 miles away a similar setup was erected. When one of these "transmitters" was excited with electricity, the other galvanometer quivered, indicating a received signal.

Unfortunately for the future of wireless, Loomis's work received only muted enthusiasm from scientists and business observers. Loomis didn't receive proper credit for his discovery until much later, after the skeptics were convinced—when no one could any longer deny the evidence that wireless telegraphy was possible.

Other nineteenth-century experiments were carried out by scientific giants such as Michael Faraday, Nichola Tesla, and Heinrich Hertz. Hertz managed to generate signals as high as 500 MHz and transmit them across short distances, proving positively that wireless telegraphy was possible. But Hertz's groundbreaking experiments were more scientific investigation than practical invention, so it was left to others to realize the art of wireless communications.

In the late 1890s, an energetic Italian inventor named Guglielmo Marconi started investigations and experiments that culminated in his achieving everlasting fame in the radio world. He first experimented with wireless over short distances of about 100 yards or less—across the garden of his father's home in Italy.

Ignored by the Italian government, Marconi took himself and his apparatus to England, where in 1896 he reached across a distance of some 2 miles over the Salisbury Plain with his wireless telegraph. Marconi rapidly increased this distance to 7 miles, and by the end of 1897 achieved distances of 10 miles between ships at sea. In 1899, Marconi accepted a suggestion of Sir Oliver Lodge and inserted a transformer device called a *Branly coherer* between the antenna and the detector. This new arrangement improved the sensitivity of the receiver and made it possible for Marconi to transmit across the English Channel—a distance of 32 miles. Finally, Marconi was ready for the big test that launched wireless into the public imagination—transatlantic telegraphy.

Transatlantic Wireless

The coast of Newfoundland in December is windswept and cold—a good place for all but natives to avoid. But it was to Newfoundland that Guglielmo Marconi went in the late fall of 1901, arriving at St. Johns on December 6 to conduct a world-changing experiment—transatlantic "wireless" (radio) telegraphy. Prior to that time, wireless had been little more than experimental, operating only over relatively short distances.

Marconi had excited the American imagination a few years earlier, in September 1899, when his wireless telegraphy apparatus was used to report the results of the International Yacht Races off the coast of New Jersey. For the first time, those ashore could have a "real-time" report of a race in progress off-shore and out of sight.

The success of this event led to demonstrations to the United States Navy in October and November 1899. Marconi installed his gear aboard the U.S.S. *New York*, an armored cruiser, the battleship U.S.S. *Massachusetts*, the U.S.S. *Porter*, a torpedo boat, and at a shore location at Highland Light, Navesink, New Jersey. The first official U.S. Navy wireless message was transmitted on November 2, 1899. It was sent from the *New York* to the shore station, requesting arrangements be made for refueling.

Despite the success of Marconi's apparatus aboard ships, ranges were still very limited. A typical shore station had a range of a few hundred miles, while shipboard transmitters were limited to as little as 40 miles and did not transmit quite as far as the shore stations under the best of conditions. It was to prove that wireless (renamed *radio* in the United States in 1910) could work over large, indeed intercontinental, distances that Marconi went to Newfoundland in 1901.

Marconi had erected a powerful spark-gap transmitter at Poldhu, Cornwall (United Kingdom), and arranged for his operators to transmit a standard message consisting of the letter *S* in Morse code at scheduled times during each day. On his windswept hill in Newfoundland, Marconi and his assistants erected the most sensitive radio apparatus then available and raised an antenna 400 feet into the sky, tethering it to a large kite. They listened for the elusive *S* transmissions. Finally, on December 24, the signal *S* was heard around noontime. An excited Marconi called his assistants to verify the reception, and then spread the news to the world.

The news of Marconi's success electrified the world. Skepticism melted and the wireless became accepted. Even today, when intercontinental radio is commonplace, Marconi's accomplishment is significant because of the crude and terribly insensitive equipment that was state of the art in 1901.

The message was clear to nearly everyone. No longer would ships at sea be out of touch for weeks at a time as they crossed the storm-tossed, iceberg-ridden North Atlantic. When mishaps occurred, the possibility of rescue was greatly enhanced once ships carried wireless gear. If the operator could send out a distress message, then other shipboard operators and shore stations could be alerted, and ships could be diverted to the scene to pick up survivors.

New Inventions

The evolution of radio advanced only slowly until the invention of the vacuum tube. In 1877, Thomas Edison had noted a strange current flowing in one of his electric light bulbs. Edison had placed a small, positively charged electrode, or *anode*, inside the bulb in an experiment to find a means for extending the useful life of his electrical lights (themselves a new invention) by decreasing the carbon emissions that coated the glass. When the lamp was turned on, a small current flowed from the light filament to the anode. This current flow was duly noted in Edison's logbook and is now called the *Edison effect*. But Thomas Edison was deeply involved in electric lighting, so he missed this potential invention.

In 1904, an Englishman named J.A. Fleming invented and patented a device based on the Edison effect. He created a two-electrode vacuum tube, or *diode*. The diode has the property of passing electrical current in only one direction, like a water valve, so it was immediately dubbed the *Fleming valve*. The nickname *valve* is still used in the U.K. to denote vacuum tubes.

Unfortunately, the Fleming valve, although useful for detecting wireless signals, was not much more effective than other nonvacuum tube detectors then on the market. It was, however, a fundamental patent and so formed the basis for a series of lawsuits in years to come. It is unfortunate that much of the early history of radio is a history of litigation between contentious giants who were too selfish to realize that cooperation would have made them all richer.

American inventor and scientist Dr. Lee DeForest started his own company with the help (it later turned out) of less than totally scrupulous financiers. Although the company seemed to prosper, it fell on hard times rather quickly because of the financial misdealings of the partners. DeForest always maintained that he was ignorant of these problems, but was not widely believed at the time.

Lee DeForest's greatest achievement was the invention of the first amplifying vacuum tube. He inserted a wire mesh grid in the space between the negative electrode, or *cathode*, and the anode of Fleming's "valve." When a signal voltage was impressed across the grid-cathode path, the signal voltage appearing across the anode-cathode was found to be magnified. *Amplification* is the control of a larger signal by a smaller signal. The DeForest *audion* tube was to prove pivotal in the invention of future radio apparatus.

Broadcasting

Although radiotelephone transmission was achieved as early as December 1906 (from the station at Brandt's Island, Mass.), and the Secretary of the Navy transmitted a message to ships at sea from the famous Navy station NAA, Arlington, Va. in 1916, it was not until 1920 that broadcasting became a reality in the United States.

Pittsburgh's KDKA, which is still on the air, grew out of radiotelephone experiments by Dr. Frank Conrad that were initiated in 1920 under the auspices of Westinghouse. On November 2, 1920, KDKA made history by broadcasting the results of the presidential election that sent Warren G. Harding to the White House. The broadcast commenced at 6:00 P.M. and lasted until well into the next day.

Development of Radio Receivers

This book is about classic radio receivers—anything from the early Branly coherers to the beginning of the solid-state era. Most of the radios that fall into this category are vacuum tube models, so it is to those that we will turn our attention first. In the rest of this chapter you will discover how the art of radio receiver design developed to its present state of sophistication.

The job of any radio receiver is to detect a radio wave and retrieve the information that it carries. It must select the desired signal from the large number of radio waves on the air, exclude others, amplify it to a useful level, and then extract whatever information that it carries (modulation). How well any given design performs these functions determines the quality of the specific receiver.

The very earliest receivers are not easily recognizable as such by today's standards. One early "receiver" was a magnetic induction coil placed in close proximity to a compass. Deflection of the compass needle provided an indication of a radio wave from a nearby spark-gap transmitter. Range was limited to only a few meters—basically from one laboratory bench to another, or as in the case of radio pioneer Guglielmo Marconi, across the backyard garden. Other early detectors included sensitive spark gaps and telegraphic electromechanical tappers.

The "holy grail" of early radio was increased transmission distance, which in practical terms meant that the questors worked on more powerful transmitters, improved transmitting and receive antennas, and more sensitive receivers. As congestion of the airwaves grew worse, the designers of radio receivers also had to address the issue of *selectivity*, as well as sensitivity. Selectivity is the ability of the radio receiver to separate stations that are close together, without undue interference between them. One early public demonstration was marred by competitors Lee DeForest and Guglielmo Marconi using the same frequency.

Transmitters in the early days of radio consisted of spark gaps (several different designs), Alexanderson alternators, vacuum tubes, and other types. Alternator transmitters were similar to the alternators that generate electricity in automobiles or electrical power plants. The frequency of alternator output is a function of the number of poles and the rotation speed. The principal difference between the power station alternator and the radio transmitter alternator is that there are many more poles and a faster rotating speed.

Vacuum tubes did not become practical for transmitters until the early 1920s, although some costly experimental models existed much earlier than 1920. Even in the 1920s, tubes were extremely expensive, produced only a low power level at RF frequencies, tended toward instability, and were short-lived.

Transmitting antennas for commercial wireless telegraphy stations (there were no radio broadcasting stations until 1920) were tall, steel towers, not unlike the towers seen at AM broadcasting stations today. For example, at Navy station NAA in Arlington, Va. (Fig. 1-1) there were one 600-foot and two 400-foot towers. Those towers were taken down just prior to World War II to accommodate Washington National Airport, which was then under construction only a short distance away.

Today, the towers still exist, and are erected at Annapolis, Md., on the Chesapeake Bay. Although the Arlington site is still used by the Defense Communications Agency (both buildings still stand), it was decommissioned as an active Navy radio station in historic ceremonies in 1956. For several years a Navy Military Affiliate Radio System (MARS) station, K4NAA, was active at the old site. Local legend has it that the foundations for the old towers are still present, but to the uninitiated look like a resurfaced parking lot.

The reason for the high towers is that the wavelength of the signal determines the size of the most efficient antennas; 1/4 wavelength and 1/2 wavelength are common. Wavelength is inversely proportional to operating frequency ($L_{meters} = 300 \times 10^6/F_{MHz}$), so at the frequencies then used for radio the towers had to be very long. For example, the 113 KHz (0.113 MHz) NAA signal has a wavelength approaching 2,700 meters, so even a 600-foot tower was an inefficient "trade-off" antenna. Commercial radio traffic was carried on wavelengths of 20,000 meters (15 KHz) to 450 meters (660 KHz), while amateur radio operators were consigned to the supposedly useless frequencies in the shortwave region (200 meters and less).

The first attempts at improving radio receivers involved finding better *detectors*. A device that will do the job of detecting radio signals will exhibit a nonlinear impedance over cyclical excursions of the applied signal. The range of telegraph

Naval Research Laboratory

1-1 The Navy's station NAA in Arlington, Va. The two 400-foot towers and one 600-foot tower was torn down in 1941 to make room for National Airport. The towers were re-erected on another site near Annapolis, Md.

clicker detectors was extremely limited—hundreds of feet at best—so alternatives were sought. Earphones, which require only a small amount of signal power to operate properly, replaced the clicker, while devices such as the Branley *coherer* (Fig. 1-2) and *liquid barretter* were used for the detector.

The coherer is now attributed to Branly, but it was originally invented in slightly different form by a Professor Hughs in 1878. Unfortunately for Branly, Marconi adopted the coherer and patented it himself—leaving Branly bitter for many years.

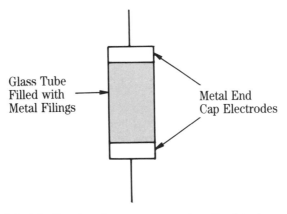

1-2 The Branly coherer was an early radio detector.

The coherer shown in Fig. 1-2 consisted of a glass tube filled with metal fillings, usually iron. The radio signals from the antenna were coupled to the coherer. The metal filings provided the nonlinear impedance needed to perform detection.

Unfortunately, the radio waves tended to magnetize the metal filings in the coherer, and that caused them to adhere together. The recovered signal would soon become weak and distorted until mechanical stimulation restored the coherer to proper operation. Consequently, a *tapper* or *decoherer* was often built into the receiver to reset the detector. When the output began to *distort*, or drop in sensitivity, the operator used the tapper to "reset" the coherer.

Crystal sets

It was soon discovered that certain natural minerals such as the bluish-gray crystalline lead ore, *galena* (lead sulphide—PbS), had a property that allowed them to detect radio signals: galena passes current in only one direction. We now know that galena forms a natural pn-junction diode at certain points on its surface.

These crystals are basically a naturally occurring semiconductor diode, and can detect both radiotelegraph and radiotelephone signals. Because galena is a cubic crystal, the galena detector was called a *crystal detector*, and radio receivers using them were called *crystal radios* or *crystal sets*.

Figure 1-3A shows the construction of a crystal detector. The galena crystal was set into a cup, and held fast by setscrews. The cup forms one electrode. Because the rectification property needed for radio detection is only found at certain spots on the surface of the crystal, a *cat's whisker* probe (which is the other electrode) was used to find the correct spot. Because it was often difficult to find the correct spot for the cat's whisker electrode, undue vibration of the radio could provoke justified hostility on the part of the operator.

The crudest form of crystal-set radio receiver is shown in Fig. 1-3B. A long wire antenna is connected to one side of the crystal, while a ground is connected to the other. A pair of high-impedance earphones is used to hear the demodulated radio signal. Even today, a very crude crystal set can be fashioned by using a germanium

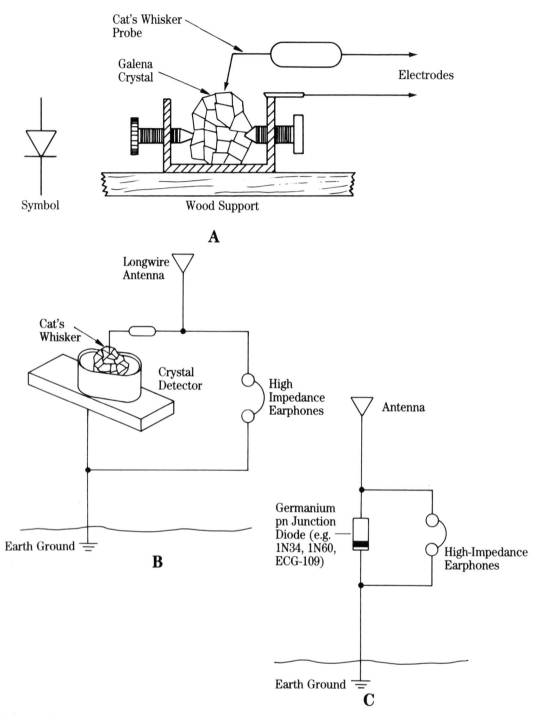

1-3 (A) Galena crystal holder for crystal sets, (B) schematic for a crystal set receiver using a galena crystal, (C) simplest crystal set today uses a pn-junction diode across a pair of high-impedance earphones.

pn-junction diode, such as the 1N34, 1N60, or ECG-109, shunted across the earphones (Fig. 1-3C).

The simple crystal set of Fig. 1-3 works only with strong, relatively local signals. Improved sensitivity was obtained with circuits such as Fig. 1-4A. In this crystal set the RF signal flowing between the antenna and ground is developed across an inductor (L). This design resulted in a stronger signal available for detection.

The crystal is connected to a sliding tap on the inductor coil. This arrangement served two basic purposes. Perhaps the most important is that the tap allows the low crystal impedance to be matched to an impedance point on the coil. Maximum power transfer always occurs in electrical systems when the source and load impedances are matched.

A further refinement (Fig. 1-4B) made the coil into an RF transformer by adding a small winding (L1) to the main coil (L2). Using transformer terminology, L1 is the primary winding and L2 is the secondary winding. The antenna and ground circuit is

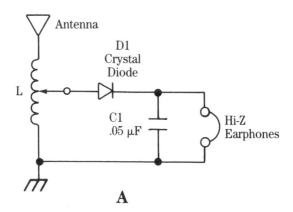

A

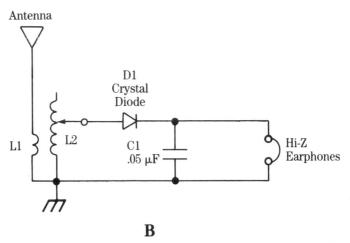

B

1-4   (A) Simple inductor "tuned" crystal set matches crystal impedance with an adjustable inductor. (B) Adding an antenna coupler winding (L1) to the crystal set coil increases output.

connected to the small coil (L1). When RF current flows in the primary winding it sets up a magnetic field that induces a signal in the secondary winding (L2). Because there are more turns in the secondary than in the primary, the voltage across the secondary is larger than the primary voltage received from the antenna.

Even in the early days of radio, the level of interference mounted rapidly, so some means was needed for selecting between signals of different frequencies. The earliest crystal sets could select desired signals and discriminate against unwanted signals only by virtue of antenna dimensions. A resonant antenna (e.g., 1/4 or 1/2 wavelength) naturally works better on its resonant frequency than at other frequencies. Nonetheless, the antenna will also pass all other frequencies to some extent.

The use of the antenna to select signals is woefully inadequate. It was found, however, that combining an inductor with a variable capacitor forms a tunable resonant circuit that could select one radio frequency, or a narrow band of frequencies, while attenuating frequencies away from the resonance point. The circuit in Fig. 1-5A is the simple version, while that in Fig. 1-5B offers an impedance matched to the detector diode.

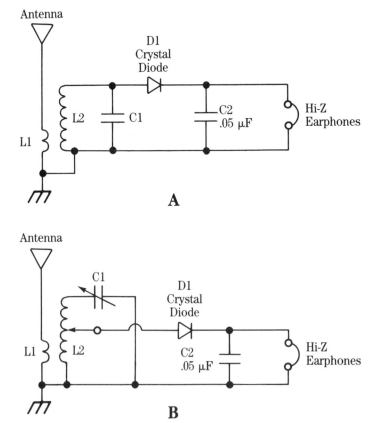

1-5 (A) A tuned crystal set allows separation of stations. (B) Combining tuning with impedance matching makes a high-performance crystal set.

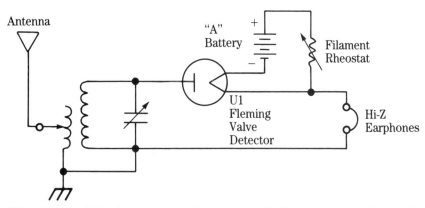

1-6 "Fleming valve" (diode vacuum tube) used as a radio detector in a receiver similar to a crystal set.

The Fleming valve was a vacuum tube diode that passed current in only one direction. The Fleming valve was therefore suited as a radio signal detector (Fig. 1-6). This circuit is essentially a crystal set, but has a vacuum diode as the detector in place of the galena crystal. A rheostat (variable resistor) in the filament circuit controlled sensitivity. Although the vacuum tube represented a step forward, it was expensive in the early years and it was not always perfectly clear to the user that the tube was superior to the natural crystal.

Although the various crystal sets were a great improvement over the coherer receivers, they still suffered badly on weak signals. In many cases, a signal was strong enough to overcome natural losses in the crystal, but was too weak to effectively drive the earphones to audible levels.

In 1906, a new invention by Lee DeForest greatly improved radio receivers. DeForest added a control grid (see chapter 3) to the "diode" (two electrode) vacuum tube previously invented by Fleming in England, who in turn based his work on a discovery made in the 1870s by Thomas Edison. The grid was a third element in the tube, so these tubes were called *triodes* (three electrodes). The advantage of the triode was that it allowed the amplification of small signals.

The earliest triodes worked only at low frequencies, but could certainly amplify audio signals. Even though the earliest tubes were inefficient and costly, the DeForest triode was often used as a post-detection audio amplifier (Fig. 1-7). The amplified audio crystal sets were a great improvement over the nonamplified versions.

Regenerative detector receivers

Another limitation of earlier designs was that both vacuum tube (diode) and galena crystal detectors were inefficient, especially on weak signals. But a Columbia University, N.Y., student (later faculty) named Edwin Armstrong invented a startlingly sensitive detector that was based on the principle of feedback. Called the *regenerative detector* (Fig. 1-8), Armstrong's circuit coupled a little of the RF amplifier output signal back to the input circuit.

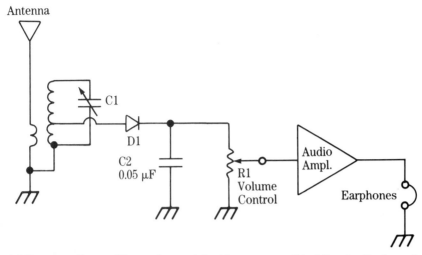

1-7   Adding an audio amplifier to the crystal set became possible following the invention of DeForest's audion.

When the degree of coupling was adjusted properly, either by a series-connected variable resistor or mechanical proximity, the circuit would be on the verge of oscillation. An input signal would force the circuit even closer to oscillation, resulting in the conditions needed for detection.

Unfortunately, the fact that maximum sensitivity occurred at the point of oscillation made regenerative receivers oscillate easily. The radio would then become a little transmitter, and cause interference to other receivers in the

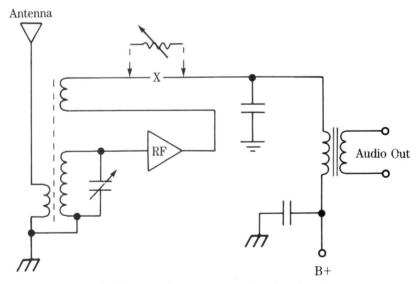

1-8   Armstrong's regenerative detector circuit.

neighborhood. Some early amateur radio transmitters were in fact Armstrong oscillators, essentially the same as a regenerative-receiver detector deliberately allowed to oscillate. By the mid-1920s regenerative radios were no longer made by the major producers. Only small producers and kit suppliers offered "regens."

Tuned radio frequency receivers

Once vacuum tubes were refined enough to amplify radio frequencies as well as audio, they were rapidly pressed into service as radio-frequency (RF) amplifiers (Fig. 1-9). If tuned LC circuits were used on at least the input of the tube (they were also used on the output of the tube circuit when instability could be controlled), then the amplifier would selectively amplify only the frequencies of interest. A receiver with such a tuned RF amplifier ahead of the detector was called a *tuned radio frequency* (TRF) receiver.

Cascading two or more RF amplifiers (Fig. 1-10) produced the classic design usually called "TRF." It offered much better sensitivity and selectivity than single-stage designs. But there was a cost to the new cascade designs. Either several tuning knobs were required (one for each variable capacitor), or all of the variable capacitors had to be "ganged" together on a common tuning shaft.

In addition, unneutralized triode RF amplifiers tend to oscillate on or near the resonant frequency of the LC tuned circuits. When such instability occurred, the oscillation signal heterodyned with the received RF signal, or with oscillations from other stages, to produce a howling shriek in the earphones. In many households only one member of the family had a sufficiently deft touch to correctly adjust the classic TRF radio. Dial settings were also unreliable, and stations often could not be relocated at the same spot on the dial.

Neutrodyne receivers

The serious instability of the TRF radio was reduced in part by a modified version of the design called the *neutrodyne* receiver. This design was patented by Professor Louis Hazeltine of The Stevens Institute of Technology. The neutrodyne design was

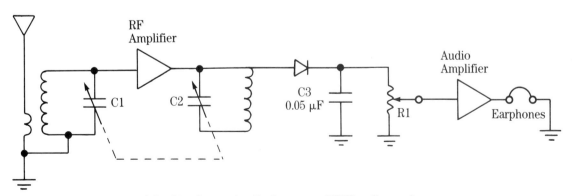

1-9   Simple tuned radio frequency (TRF) radio receiver.

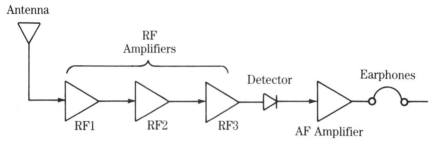

1-10   Multistage TRF radio receiver.

based on the fact that improved stability resulted when a small amount of *degenerative*, or negative, feedback was introduced to counter the *regenerative*, or positive, feedback that was inherent in the TRF design. The positive feedback results from interelectrode capacitance within the vacuum tube, as well as unintended coupling external to the tube.

Hazeltine's neutrodyne receivers proved very stable, and for the first time dial calibrations had practical meaning even on low-cost receivers. Station dial settings were actually repeatable. Between 1924 and 1927, companies such as F.A.D. Andrea (with the Model FADA-160) and Freed-Eisemann (with the Model NR-5) produced and sold a large number of neutrodyne receivers.

When audio tubes increased in power level, the radio truly became a family entertainment center because a loudspeaker could replace the earphones, allowing the entire family to hear the broadcast at the same time. Unfortunately, only a few people could afford radio receivers. At $300 to $700, models cost several months' salary for a typical 1920 worker. In 1919, vacuum-tube radios carried price tags on a par with small houses, but within a few years they were more affordable. By the 1930s, radios were within the reach of large numbers of families as the mass market era began.

Second-generation TRF radios

Starting about 1927, radio tube manufacturers offered tubes with a fourth element called a *screen grid* (see chapter 3). Much of the instability (oscillations) seen in triode-tube circuits is attributable to interelectrode capacitance between the cathode-grid and grid-plate elements. By placing a screen grid between the control grid and anode, tube markers effectively reduced interelectrode capacitance to a point where it could almost be ignored.

The tetrode tube led to a type of radio design called the *second-generation TRF*. These sets were produced between 1927 and 1932, and were a relatively low-cost alternative to the expensive superheterodyne radios also on the market at that time.

Superheterodyne receivers

The new TRF still suffered at least somewhat from the howls and shrieks of other TRFs. In 1920 Professor Edwin Armstrong again came up with an invention that solved the problem. The new Armstrong receiver, dubbed the *superheterodyne* (the "super" part probably came from the marketers rather than the engineers), is still the design of choice today. Although a single integrated circuit (IC) has replaced the several vacuum tubes used in the 1920s designs, the essential principle is the same. The first commercial "superhets" were produced by RCA in 1924 (Model AR-812), and later as the *Radiola Super VIII.*

A story that circulated among radio buffs concerning the superhet claims that Armstrong, who was then affiliated with Columbia University in New York City, invited David Sarnoff to his laboratory one evening to see a new radio. The superhet was so sensitive and selective that Sarnoff—a visionary and genius himself, among other things that are said of him—almost salivated over the sealed wooden box. According to this story, Armstrong would not reveal the secret of the box until Sarnoff paid for it.

The principal defect of TRF designs was stability of the RF amplifiers, especially high-gain cascade amplifier chains. It was this problem that the superhet solved nicely. Even though one can build such amplifiers for one frequency and achieve good stability, variable-frequency RF amplifiers would almost always oscillate, at least on some frequencies.

The variable-frequency TRF radio was either very complicated or very unstable. Armstrong's idea, and the basis for the superhet design, was that he redistributed the amplification and selectivity to one frequency, called the *intermediate frequency,* or IF. All received frequencies were converted to this single IF frequency, where a very stable high-gain amplifier could process them.

Figure 1-11 shows the basic superhet block diagram. The key to superhet operation is the conversion of the RF signal (frequency RF) to the intermediate frequency (IF). This job was done by nonlinear mixing of the RF signal with a signal from a local oscillator, operating on frequency LO. Frequency LO can be above

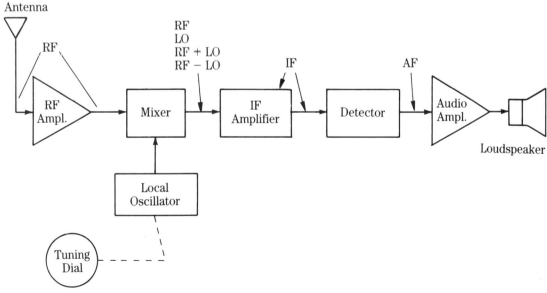

1-11 Superheterodyne radio receiver.

frequency RF or below it. The oscillator signal is produced inside the receiver. The main tuning-dial of the superhet receiver is connected to the capacitor that tunes the local oscillator.

When two frequencies such as RF and LO are mixed together nonlinearly, the output of the mixer will contain at least four frequencies: RF, LO, RF+LO (sum), and RF−LO (difference). The IF can be either the sum or the difference frequency, although for practical reasons at the time, only the difference frequency from the mixer was used. Thus, the *mixer-LO* stage, sometimes called either the *first detector* or *converter* stage, converts the RF frequency of the incoming signal to the lower IF frequency, which equals RF-LO.

Many superheterodyne receivers used an RF amplifier ahead of the mixer. This stage was sometimes called the *preselector* stage. It served a couple of different functions, even though it provided only a small amount of gain. First, it tended to unilateralize the receiver; that is, it allowed signal to flow only from the antenna into the radio. Signals inside the radio could not flow back out to the antenna where they could radiate and cause interference with nearby receivers. The particular offending internal signal that needed such unilateralization was the very local oscillator that made the superhet work.

Second, the preselector served to eliminate strong out-of-band signals that could interfere with the desired RF signal. Although the principal selectivity function was in the IF amplifier stage(s), the IF selectivity did nothing to improve the *image response* of the receiver.

The image frequency in a superheterodyne receiver is a normal (albeit unwanted) response that must be suppressed. There are at least two manifestations of image response that cause problems. First, there is the response to unwanted strong signals. Figure 1-12A illustrates the problem.

Normally, the LO frequency is spaced away from the received RF signal by an amount equal to the IF. That's how we get IF = RF−LO. In the example of Fig. 1-12A, the LO is on the high side of the RF frequency. If there is another signal present at a distance equal to the IF on the other side of the LO, then it too produces a difference frequency equal to the IF, and will pass right through the IF amplifier as a valid signal. The preselector improves rejection of these image frequencies by sharply attenuating them ahead of the mixer.

The other manifestation of image response problems is that the same signal might appear on the dial at two different points. Consider Fig. 1-12B. Initially, the radio is tuned to RF_1, so the local oscillator will be at LO_1, which is equal to RF+IF. Since a signal is present at RF_1, the receiver detects a station and everyone is happy.

But later the receiver is retuned away from RF_1 to a new frequency, RF_2. If RF_2 is $2 \times$ IF away from RF_1, now the new LO frequency (LO_2) is still displaced from RF_1 by the frequency of the IF, so a "valid" signal is produced for the IF amplifier. Again, the answer is to preselect the RF signal so as to attenuate all but valid signals.

The IF amplifier provides the bulk of the overall receiver gain, as well as the best selectivity. It is relatively easy to provide these features in a single-frequency amplifier, especially when the frequency is low. Thus, the superhet radio reached the pinnacle of performance available to any particular class of radio receiver. The

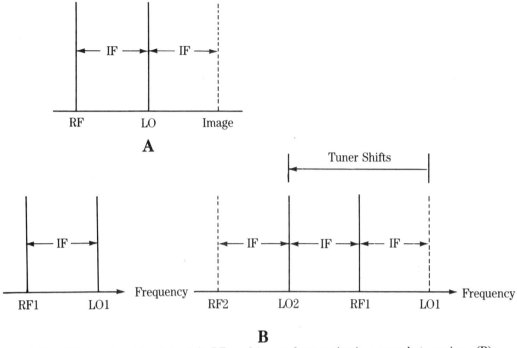

1-12 (A) Relationship of the LO, RF, and image frequencies in a superhet receiver. (B) tunable images.

design was so successful for its time that it survives even today. Except for a few special-purpose receivers and hobbyist projects, radio and television receivers even today are superhets. In the chapters to follow you will learn more about these radios, how they operate, and how to troubleshoot and repair them.

2

Vacuum Tube Radios

IN 1916 AN ENTHUSIASTIC YOUNG EMPLOYEE OF AMERICAN MARCONI WIRELESS WROTE a now-famous memorandum to his supervisor concerning the future of radio as a home entertainment medium. Four years before Frank Conrad of Westinghouse made the first official broadcast from KDKA, Pittsburgh, a visionary young David Sarnoff saw the huge potential of bringing news, concerts, and other entertainment into people's homes via the airwaves.

Sarnoff, who went on to found Radio Corporation of America (RCA) and guide it to become a corporate giant, foresaw the then huge $75 million potential of this market. But, because radio was in its infancy, and most radio transmitters were spark gaps, and therefore unsuitable for broadcasting, few people took him seriously. Although Sarnoff was right, his vision had to await the development of lower-cost vacuum tubes in the early 1920s.

David Sarnoff was wrong, by the way. His 1916 prediction of $75 million was off by a factor of ten. By the beginning of World War II radio sales approached $700 million.

In this chapter we take a brief look at early vacuum tube radios, and trace their general development as a commercial product from the 1920s to the later 1940s. Radio collectors may wish to go deeper into this subject than we have space for in this chapter. For those readers we recommend: *The Radio Collector's Directory and Price Guide* by Robert E. Grinder and George H. Fathauer (Ironwood Press, Box 8464, Scottsdale, AZ 85252).

Grinder and Fathauer give guidelines on identifying the approximate age of radios. In addition, their price guide lists a large number of radio receivers by model, and the first year of manufacture for each. The prices shown, however, are only guidelines because actual prices change, according to area, circumstance of purchase, and the time between publication and actual purchase.

Radio Construction Styles

Radio construction methods evolved over the years, and as a result, specific styles of construction can be used to date the radio to within a few years. Prior to 1920, radios resembled scientific apparatus, which indeed they were. About 1920, the *breadboard* type of construction (Fig. 2-1) was popular. A large rectangular board was used as the base of the receiver, and the components were mounted on top. Point-to-point wiring was found on the underside of the breadboard. Three sets of tuning components were provided; one for each RF circuit. Ganging controls had not yet been invented.

Breadboard construction was not altogether satisfactory for home radios once they were intended for other than the uncritical enthusiast. As broadcasting became a viable medium, radio design became oriented more towards the entire family. Some of the first low-profile table models (Fig. 2-2) appeared in the early to mid-1920s.

Like breadboard radios, many of the early table models used two or three separate tuning controls. By the end of the 1920s, however, the "three dialers" were disappearing and more modern forms of radio with a single control and window dial were coming on the scene. Many of these radios used an external speaker in the form of a horn.

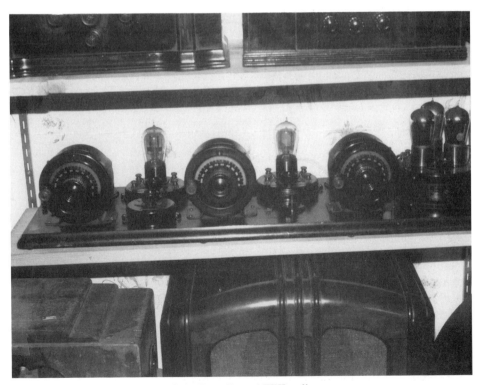

2-1 Breadboard TRF radio.

2-2   Collection of antique TRF radios at a collector's swap-meet.

Probably the one radio design that is truly representative of radios in the early period is the *Cathedral* model shown in Fig. 2-3. These models were also known as the *Gothic* or *upright* radio. The cathedral radio was typically 18 to 22 inches tall, and had a base between 12 and 16 inches square. Tuning was generally of the one-knob type, and the speaker was integral to the radio cabinet. These radios were made from the late 1920s until the early 1930s.

Another popular American radio design was the *Tombstone Upright* shown in Fig. 2-4. These radios were approximately the same size as the Cathedral, or possibly a little larger. Unlike the Cathedral, however, the Tombstone radio was more squared off, and a bit less costly to manufacture, I'll bet. These radios were produced from the early- to mid-1930s.

In Britain, radio manufacturers were producing their own radio designs. What the Tombstone and Cathedral models were to American purchasers, the round model of Fig. 2-5 was to the British user. Companies like Eddystone and Pye made radios in Britain paralleling American developments, and in some ways exceeding them.

Also becoming popular in the late 1930s and 1940s were the "modern" console, or floor model radios (Fig. 2-6). Large consoles were made in the 1920s, but were extremely expensive, and limited to a more affluent market. But more mass-market consoles began to appear in the late 1930s, and the radio became an everyday piece of furniture.

2-3 Cathedral-style radio.

2-4 Tombstone-style radio.

2-5   Round radio typical of British models pre-WWII.

2-6   Zenith console model.

2-7   Mantlepiece radio.

One of the wonders of electronics over the years has been miniaturization. And, although by today's standards nothing produced in the 1930s and 1940s was "miniature," to the buyer of the late 1930s the small "mantlepiece" and midget table models (Fig. 2-7) were a triumph or contemporary design. In the late 1940s and early 1950s the table model was still popular, and often sported an FM band as well (Fig. 2-8).

Early Vacuum Tubes

Vacuum tubes were manufactured from the first decade of the twentieth century, but did not become firmly established as a viable industry until the post-World War I era. Even then, vacuum tubes were terribly expensive—the cheapest of them cost a week's salary.

The late McIvor Parker (W4II), who was involved in radio from the mid-1920s, once related to me how he saved for months to buy three vacuum tubes at $7.50 each, in order to build a ham radio receiver. A careless slip with a screwdriver shorted B+ to the filaments line, and burned them out instantly. To a young radio enthusiast that fatal slip was a tragic event and a financial disaster.

The tube complement of a radio receiver is often used to establish the approximate date of manufacture of the set. One of the best guides to this is laid out in *The Radio Collector's Directory and Price Guide*, in which a chart of tubes and their dates is presented.

2-8   Table model radio.

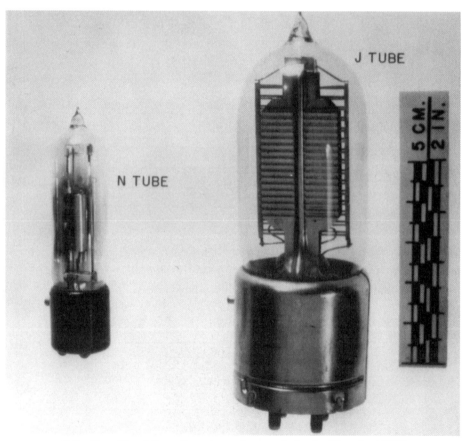

2-9   Type-N and Type-J receiver tubes made for the Navy by AT&T.

Figure 2-9 shows several different types of very early vacuum tubes. At the far right is the DeForest *Audion* tube. These proved unreliable and were expensive. The Type-N and Type-J tubes also shown in Fig. 2-9 were made by AT&T for use in U.S. Navy receivers. The Type-J incorporated an oxide-coated cathode (1913), and was eventually miniaturized into the Type-N (1920).

AUDION

2-10 DeForest audion tube.

2-11   Tubes typical of the 1920s.

Tubes in the 1920s generally had four-pin bases, and envelopes such as shown in Fig. 2-11. Typical type numbers included 200A, 201A, WD-11, and WD-99. By 1930 the first digit of the tube-type number was generally dropped, so type numbers then current that included tubes such as the '30, '37, and '45 were common. Many of these late 1920s and early 1930s radio tubes had five-pin bases and glass envelopes, and starting in about 1932 six- and seven-pin types became available. Some also included a grid cap on the top of the tube (Fig. 2-12).

2-12   Tubes typical of the early 1930s.

2-13   Octal tubes (mid-1930s).

The late 1930s saw the development of eight-pin metal-can octal tubes such as Figs. 2-13 and 2-14. These tubes used a special base that had a plastic prong in the center that contained a keyway ridge. This feature allowed the tube to be inserted in only one direction. Previous tube bases were keyed by making two of the pins larger than the rest of the pins.

2-14   Octal tube.

2-15   Miniature seven-pin and nine-pin glass tubes.

Also available in that era was the eight-pin *Loctal* socket. Because the Loctal tube has the ability to lock into its socket, these tubes were used extensively in World War II radio receivers.

Just prior to World War II, the glass miniature tube was invented, with the seven-pin type coming first and the nine-pin type a little later. These tubes (Fig. 2-15) were the standard for vacuum tubes right into the transistor era, when tubes of all sorts began to fade from the scene. Although the *Nuvistor*, the *Lighthouse*, and the *Acorn* tubes were also invented, they saw limited use or no use in consumer radio receivers.

3

Vacuum Tubes— How They Work

VACUUM TUBE, OR *ELECTRON TUBE*, AMPLIFIERS HAVE BEEN WITH US FOR DECADES, and in fact were the first amplifying electron devices. From around the turn of the century until only a few years ago, *electronics*, as opposed to *electricity*, meant vacuum tube circuits. Since the transistor and integrated circuit have come of age, however, the vacuum tube has faded into the background so much that electronics professionals who received their training in current technical schools do not even study the vacuum tube. This chapter on how vacuum tubes work is included because antique radios and classic radios used vacuum tube technology.

If some engineers of thirty years ago seemed a little slow to appreciate the then new-fangled transistor, radio designers in their grandfathers' day were equally excited about vacuum tubes—if they could make the darn things work with anything like reliability. Even though the tubes presented demonstrable advantages over existing equipment, they were not too reliable in the early days.

Certain vacuum tube principles were, however, known in the latter part of the previous century. Indeed, physicists used evacuated glass tubes for electrical experiments for many years in the nineteenth century.

Edison Effect

When Thomas Alva Edison was conducting some of his many electric lamp experiments in the last quarter of the nineteenth century, he observed and recorded a number of interesting phenomena. One of these, which bears his name, is the *Edison effect*. This phenomenon was discovered during experiments that were designed to find means to eliminate, or at least reduce, the carbon blackening of the insides of Edison's electric lamps.

This blackening reduced the light output of the bulbs prematurely. One method involved placing an extra electrode inside of the lamp's glass bulb along with the filament. Edison noted a small current flow to that electrode whenever a positive electrical potential was applied to it. Since a positive potential attracts electrons, it may be assumed that a source of free electrons must somehow be connected with the glowing filament.

Thermionic Emission of Electrons

The source of the free electrons inside of a lamp is the electrically heated incandescent filament wire. When an electric current is passed through a conductor, a certain amount of heat is generated. In the case of the lamp filament, this heating is so intense that light from red to white is given off. The color of the light is dependent upon the amount of current flowing. When an electrical conductor is heated, its free electrons take on additional kinetic energy. This follows, in fact, from one of the physical definitions of temperature. Increased kinetic energy means that the velocity of the individual electrons is increased.

Imagine the situation inside such a metallic conductor. The free electrons will be moving rapidly, and in all directions at once. They will, of course, bump into each other, creating millions of collisions per second. Out of all of these helter-skelter electrons, only a few percent will have both the direction of travel and the velocity to escape when they are near the hot surface of the filament wire.

The electrons that escape from the filament surface form a *space charge* (see Fig. 3-1) surrounding the cathode in the glass envelope. It was the space charge electrons that were attracted to the anode to form the *Edison effect current*.

The name of this phenomenon is the *thermionic emission of electrons*, i.e., where surface electrons on the wire boil off into space because of the conductor being heated. Thermionic emission, then, is responsible for the current on which vacuum-tube operation depends.

The Vacuum Tube Diode

J.A. Fleming first used the Edison effect to produce a diode (*di*—two, *ode*—electrode), or two-electrode, vacuum tube. Fleming's tube used a *cathode* as a source of thermionic electrons, and Edison's extra electrode as an electron collector. When positively biased with respect to the filament (as in Fig. 3-2A), this anode will attract and collect the space-charge electrons, and a current will be registered on the milliammeter.

When battery V1 is reversed, however, the anode is negative with respect to the filament. This negative charge on the anode repels the negatively charged electrons produced by the cathode, so they are driven back towards the filament. In this case, no current will be registered on the milliammeter—it will read zero. This tells us that Fleming's diode will pass electrical current in only one direction. It may be a little difficult to see the usefulness of this at this point, but it proves to be a highly useful phenomenon. But more of that later (see also chapter 5).

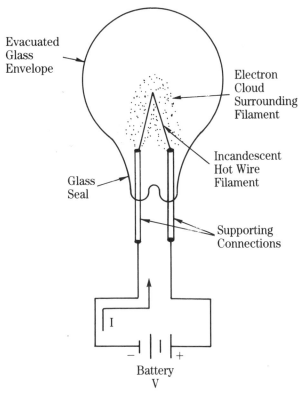

Evacuated Glass Envelope

Electron Cloud Surrounding Filament

Glass Seal

Incandescent Hot Wire Filament

Supporting Connections

I

Battery V

3-1 Thermionic emission of electrons from a hot filament in an evacuated glass bulb.

Cathodes

In our simple example the cathode was an incandescent (i.e., "gives off light when heated") filament. This is an example of a *directly heated cathode*. Figure 3-3A shows both the construction and schematic circuit symbol for the directly heated cathode.

An indirectly heated cathode is shown in Fig. 3-3B. The physical construction is shown as a cutaway drawing. The schematic symbol is also shown.

Any cathode material will emit electrons when heated. From a theoretical point of view, it does not matter where the heat comes from. You could, for example, extend the cathode out of the glass vacuum-tube envelope so that it may be heated with a cigarette lighter, but that approach is totally impractical.

In real vacuum tubes the heat is generated by an electrical current in an incandescent filament. Keep in mind that the current in the filament is used solely to *heat* the cathode. It does not supply the anode electrons; any heat source would create the anode current from free electrons within the cathode material.

In Fig. 3-3B we have the cathode heated indirectly from an incandescent filament. The actual cathode structure is a hollow cylinder of metal. In all but the earliest tubes the cathode is made of *thoriated tungsten*, or tungsten coated with

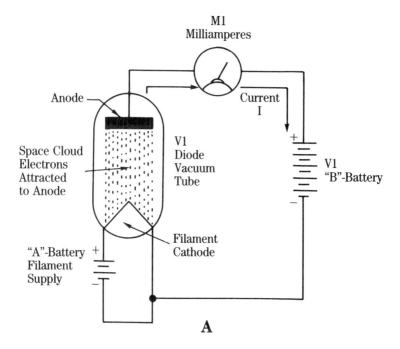

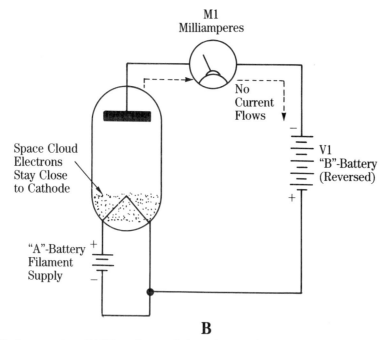

3-2   Diode operation: (A) When the anode is positive with respect to the cathode, electrons from the space cloud flow to the anode. (B) When the polarity is reversed, and the anode is negative with respect to the cathode, electrons are repelled, so current flow ceases.

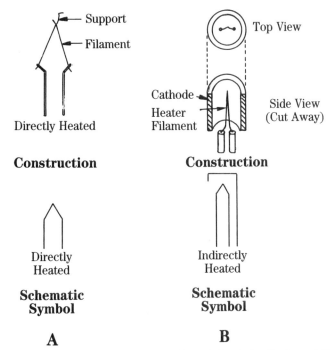

3-3   Construction and circuit symbols for (A) a directly heated cathode, and (B) an indirectly heated cathode.

thorium oxide. Inside this structure is a filament not unlike those described earlier. It is the incandescence of this filament that causes the red to white hot glow inside the tube.

Diode construction

Figure 3-4 shows the circuit symbols for both directly heated cathode (Fig. 3-4A) and indirectly heated cathode (Fig 3-4B) diodes. The diode consists of the cathode and anode inside a common evacuated glass envelope. The construction of these two diodes is shown in Fig. 3-5. The directly heated case is shown in Fig. 3-5A. Note the

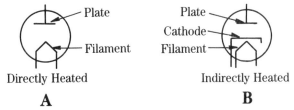

3-4   Diode vacuum tubes with (A) a directly heated cathode, and (B) an indirectly heated cathode.

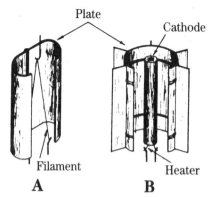

3-5   Cutaway views of (A) directly heated and (B) indirectly heated diode tubes.

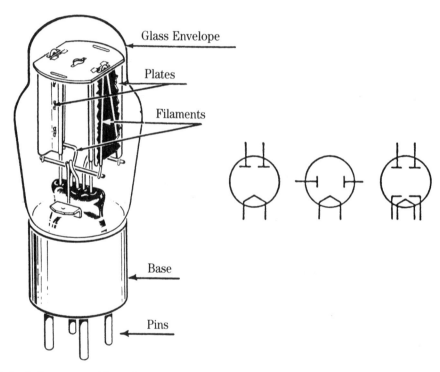

3-6   Full-wave rectifier tube construction and circuit symbols. These tubes contain two diodes in the same glass envelope, and are used in dc power supplies.

elliptical plate (anode) shape. The indirectly heated cathode variety (Fig. 3-5B) shows the cylindrical plate design.

   The construction of a dual diode power-supply rectifier tube is shown in Fig. 3-6 along with several popular alternative circuit symbols. These tubes have two separate anodes over either a common cathode, or separate cathodes over a common heater filament. The most common use of these tubes is in full-wave rectified dc power supplies.

Diode operation

Consider a circuit such as that in Fig. 3-7A. You may recognize that as being similar to Fig. 3-2, except that supply voltage V_b is variable from 0 Vdc to some positive potential. Figure 3-7B shows the anode current-vs.-anode voltage curve for a typical diode. Note that the plate current is zero when the plate voltage is also zero. It is also true that plate current is zero when the potential on the plate is negative with respect to the cathode. Current will register on meter M1 only when V_b is positive with respect to the cathode.

Before proceeding further, let us consider some basic terminology. In the discussion of vacuum-tube circuits, several different voltages are mentioned with some regularity. These have been traditionally given letters to designate their usage. The *A* voltage is the filament supply. It can range from about 1.5 volts to over 117 volts, but 2.5 volts, 6 volts, and 12 volts are probably the most common.

The electrical potential connected between the plate and cathode is usually called the *B* supply voltage. It ranges from a low of around 12 Vdc, in hybrid (or "transistor powered") car radios of the late 1950s, to several kilovolts in high-powered radio transmitters. In a short while, we shall introduce another electrode, called a *grid*. The voltage applied to the grid is usually designated the *C* supply.

The graph in Fig. 3-7B is the relationship I_p vs. V_b. In the upper ranges, this curve is what we might expect—a straight line. The *plate resistance* is the quantity:

$$R_p = \frac{V_b}{I_p}$$

Notice, however, that the plate current is not linear in the low range, where plate voltage becomes low. At these potentials only those electrons which have the greatest kinetic energy will reach the plate. At slightly higher potentials electrons with lower kinetic energy levels will reach the plate. At about 25 volts, however, all emitted electrons will reach the plate, and the curve straightens out. In this region the diode is almost perfectly linear.

Figure 3-8A shows a diode that has the B battery replaced by a transformer and ac generator. The transformer is connected so that the secondary voltage is impressed across the diode's anode and cathode. The graph in Fig. 3-8B shows the voltage applied to the anode, with respect to the cathode. Also shown is the plate current existing at these various plate voltages. Whenever the plate voltage is positive, plate current will flow. Also, note that the shape of the current curve is very nearly the same as that of the plate voltage.

On the other hand, on negative peaks of the ac voltage applied to the plate, no flow of current exists. This action follows from the diode passing current in only one direction, and is called *rectification*.

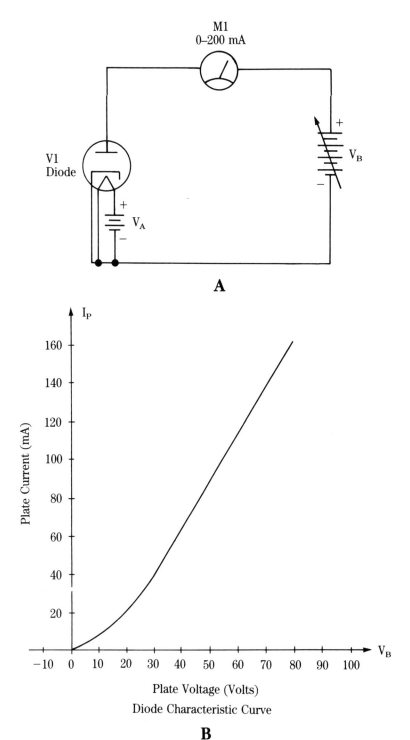

A

Diode Characteristic Curve

B

3-7   (A) A dc diode test circuit. (B) I_p-vs.-V_B curve.

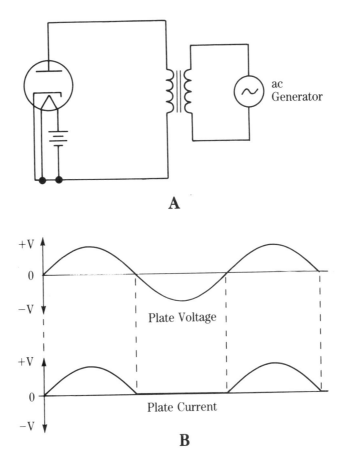

3-8 (A) An ac diode test circuit, (B) input and output waveforms.

Triode Vacuum Tubes

The range of applications enjoyed by the diode is severely limited because it cannot amplify signals. One definition of amplification is the *control of a large current or voltage by a smaller voltage or current*. Lee DeForest first devised a practical amplifying vacuum tube by inserting a control grid into the space between the cathode and anode of Flemming's diode. This new type of tube had three electrodes, or elements, as they are sometimes called; so it is called a *triode*.

Figure 3-9A shows a cutaway view of a triode vacuum-tube, while Fig. 3-9B shows the usual schematic circuit symbol. The grid is a control element, and is therefore often called the *control grid*. It could be an actual mesh, but is most often a grid of fine wires (Fig. 3-10). The two popular forms of grid are the *elliptical helix* and *ladder*.

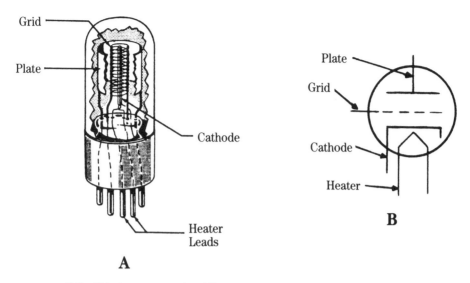

A

B

3-9   Triode vacuum tube: (A) cutaway view, (B) circuit symbol.

The grid may be viewed as an electrically porous element placed in the path of electrons flowing between the plate and the cathode. The proper modern schematic symbol is shown in Fig. 3-10B, while the most common original symbols are shown in Fig. 3-11. In the rest of this book, the modern symbol will be used even when depicting antique radio circuits.

Let us assume, first, that no voltage is applied to the grid. In that case, electrons attracted to the high positive potential on the plate will pass right through spaces in the grid to strike the plate in the normal manner. Next, let us consider a case where the grid voltage V_c is not zero.

A   **B**

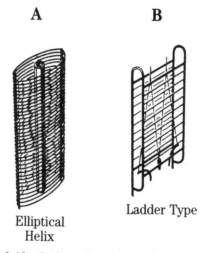

Elliptical
Helix

Ladder Type

3-10   Grid construction methods.

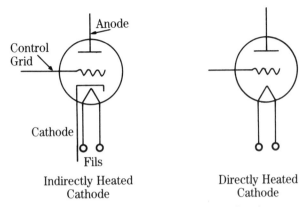

Control Grid

Anode

Cathode

Fils

Indirectly Heated
Cathode

Directly Heated
Cathode

3-11 Obsolete circuit symbols for triode tubes.

A triode circuit in which $V_c < 0$ is shown in Fig. 3-12. The grid, in this case, is negatively biased, so it will oppose the flow of negatively charged electrons from cathode to anode. At low negative values of grid voltage only a small retarding effect is apparent on the electron stream. As V_c becomes more negative, however, fewer and fewer electrons make it to the plate. At some particular value of V_c, no electrons will pass to the plate, and the triode is cut off. At $V_c = 0$, the triode tube acts very nearly like the diode discussed earlier. Positive values of V_c, however, will increase plate current to a point where the tube may well be destroyed.

Now let us consider how a triode is able to amplify. Recall that amplification means that we want to be able to control a large voltage with a smaller voltage. The circuit in Fig. 3-13 shows a triode connected as a voltage amplifier. Resistor R_L is the load used to develop the output signal voltage. Voltage V_c is the dc grid bias, while voltage V_g is the ac input signal voltage. The total voltage applied to the grid will be the algebraic sum of V_c and V_g.

If V_g is zero, then V_c is the total control over the plate current. The value of plate current I_p under this condition will be a constant static current. Alternating current

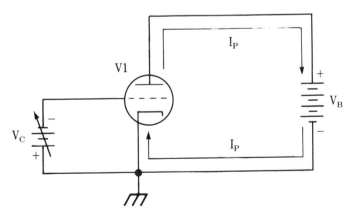

3-12 Static dc triode test circuit.

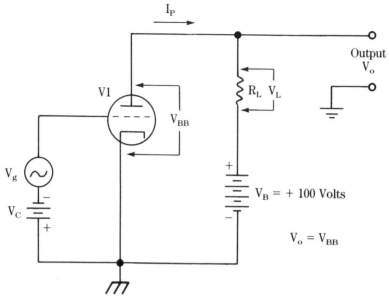

3-13   Dynamic triode circuit.

voltage V_g is in series with V_c, so it will either add or subtract its value to that of V_c, depending upon polarity. The voltage applied to the grid of V1, then, is $-V_c \pm V_g$.

The bias voltage creates a certain value of current I_p flowing in the plate circuit. This current also flows in the load resistor R_L, causing voltage drop V_L to appear across the load resistor. The actual voltage seen by the triode's plate-cathode circuit, then, will be something less than V_b. It will actually be:

$$V_{bb} = V_b - V_L$$

Grid bias voltage V_c is set so that plate current I_p is in the middle of its straight-line portion. This is the region of *linear operation*.

If V_g is allowed to swing positive, then the total grid voltage will become less negative. Plate current I_p, then, must increase, thereby increasing V_L. Under this circumstance V_o and V_{bb} will decrease. This tells us that the signal voltage output waveform will be decreasing, or "negative going."

Voltage V_o, then, becomes lower as V_g goes positive, and will become greater as V_g goes negative. Although the output waveform V_o is thus inverted (it is 180 degrees out of phase), it has the same shape and is much larger than V_g.

Voltage V_g can increase in a positive direction until it either exceeds $-V_c$, or until V_{bb} drops to zero. In the first case, if V_g overcomes $-V_c$ so much that the voltage applied to the grid is positive, the plate current I_p increases sharply. In that case, the shape of the output waveform no longer resembles the input waveform, and the output is distorted.

Likewise, distortion also occurs if V_{bb} is allowed to drop to zero before V_g reaches its positive peak. In that case the tube is said to be saturated. No further increase grid voltage will cause an increase in plate current. In either case, the output will be distorted, and the amplifier is said to be operating in a nonlinear manner.

There are certain limitations and constraints placed on vacuum tube operation, especially if linearity—the absence of distortion—is a desirable factor. The value of the grid bias voltage V_c, for example, must be set so that it is approximately in the center of the linear, or straight-line, portion of the characteristic curve (see Fig. 3-14).

The input voltage may be allowed to swing positive and negative only enough to prevent V_o from trying to exceed V_b or from dropping to a point where the curve is not linear. V_o cannot exceed V_b, but when it tries, distortion results. In either region bad distortion results. These limitations make it highly desirable to keep the tube operating only between points A and C in Fig. 3-14.

Tetrode Vacuum Tubes

High amplification in triodes requires the placement of the control grid very close to the cathode. This, unfortunately, produces a few problems. One is the fact that close proximity of the two electrodes increases the probability of them accidentally shorting together. It also tends to reduce the voltage level that may exist between them without arcing. Another problem is that close spacing tends to increase *interelectrode capacitance*, thereby deteriorating the tube's performance at high frequencies. It must also be noted that close grid-cathode spacing requires much higher plate current.

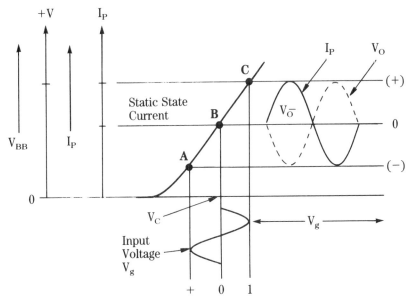

3-14 Graphic depiction of triode operation and dynamics.

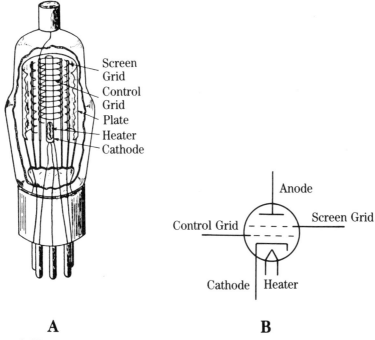

Screen
Grid
Control
Grid
Plate
Heater
Cathode

Anode

Screen Grid

Control Grid

Cathode | Heater

A   **B**

3-15   Tetrode vacuum tube: (A) cutaway view, (B) circuit symbol.

These problems can be partially overcome by the four-element, or *tetrode*, vacuum tube (Fig. 3-15). In this design, a second, but positively biased, screen grid is placed in the space between the control grid and the plate. The advantage of this construction include higher amplification at a given plate current and lower interelectrode capacitance.

In multigrid tubes such as the tetrode, the control grid is often called *grid no. 1* (G1), and the screen grid is called *grid no. 2* (G2).

In the normal tetrode circuit (Fig. 3-16), the plate and control grid voltages are

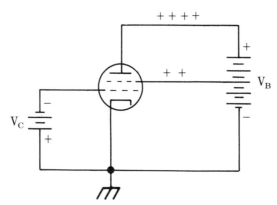

$+ + + +$

$+ + $

V_B

V_C

3-16   Normal circuit for the tetrode tube.

much the same as they were in the triode. The plate is made positive with respect to the cathode, while the control grid is biased negative with respect to the control grid.

The screen grid, placed between the control grid and the plate, is biased with a positive potential, but usually less so than the plate. Electrons that pass through the control grid are attracted by the positive potential on the screen grid, so they are accelerated as they head for the plate. Most of the electrons pass through the screen grid and continue on to the anode.

Pentode Vacuum Tubes

The five-element *pentode* vacuum tube (Fig. 3-17) was invented in response to certain limitations and problems in both triode and tetrode designs. It seems that electrons attracted to the plate are accelerated through quite a large electrical potential; therefore they acquire a large kinetic energy proportional to the combined effects of plate and screen grid potentials.

This acceleration is even greater in tetrodes than in triodes, because of the positively charged screen grid. When electrons acquire sufficient kinetic energy, they are capable of striking the plate with such force that a few electrons in the plate material are dislodged into the space between the plate and the screen grid. This phenomenon is called *secondary emission of electrons*. Some of these electrons have sufficient kinetic energy to be impelled into the region where they can be attracted by the positively charged screen grid.

Of course, this current constitutes an undesirable reverse current flow to the screen grid. Other electrons have less kinetic energy after emission from the plate, so they remain near the plate and are attracted back to the plate by its positive potential. This creates a slight variation in plate current that is not caused by a corresponding variation in grid voltage (another form of distortion).

In addition to the nonlinearity caused by secondary emission, there is also the possibility that the reverse screen current will cause the total screen current to exceed the ratings of the tube, thereby causing burnout of the device.

The problem of secondary emission is reduced in severity by placement of a third grade electrode between the screen grid and the anode, and giving it either a

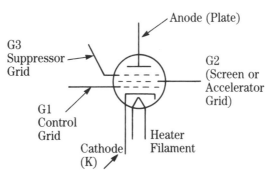

3-17 Pentode vacuum tube circuit symbol.

slight negative or zero bias. Electrons displaced by secondary emission are typically much lower in kinetic energy than the primary cathode electrons.

Secondary electrons, therefore, can be repelled back towards the plate by a tiny negative voltage on the third grid, called a *suppressor grid* because it suppresses secondary emission, and so are forced back to the plate. Primary electrons, on the other hand, have sufficient kinetic energy to overcome the weak negative field of the suppressor as they pass from cathode to plate; so they will continue on through the suppressor grid and strike the plate.

The suppressor grid may be biased at either a zero volts potential, or it may be given some specific small negative bias. In many circuits the suppressor grid is tied to the cathode, so the two elements will have the same bias. In fact, there are many vacuum tubes in which the suppressor is internally tied to the cathode.

Several factors make the pentode desirable for a wide range of applications. One is the particularly high amplification factor evident in the pentode design. Most pentodes have amplification factors much higher than triode and tetrode tubes operated at similar plate potentials. This allows a great deal of voltage gain in a relatively small package.

Another valuable benefit is the extremely low values of interelectrode capacitance present in pentodes. These are made lower by the fact that the control grid "sees" several capacitances effectively in series, which makes the total capacitance lower. In many circuits the tube's interelectrode capacitance is so high that external circuitry is required to compensate for its effects. This technique is called *neutralization*. It is needed primarily in RF circuits. In many circuits, pentodes will operate satisfactorily without neutralization, whereas a triode or tetrode would not.

On the debit side, however, is the fact that the pentode plate resistance tends to be very high: values between 50 Kohms and 1 Megohm are common. This high impedance makes the pentode a little more difficult to impedance-match, almost forcing the use of transformer or RC-coupling techniques between the tube plate circuit and the load.

Figure 3-18 shows two sets of vacuum tube characteristic curves from the type 6AN8A vacuum tube. This tube was selected for our example because it is actually a

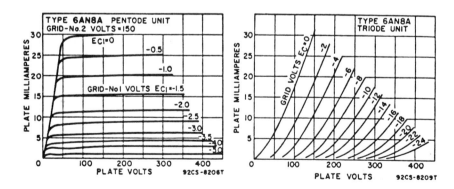

3-18   6AN8 curves.

dual tube that has a triode and a pentode in the same glass envelope. This example affords us the opportunity to study both types, in a setting where we may more easily appreciate their respective differences.

Notice that each family of curves actually contains several characteristic curves for each tube. These curves are taken at different values of grid-bias voltage. Note that the V_c values for the triode are much higher than those in the case of the pentode. This is a consequence of the lower gain factor for the triode.

If you examine the graph of plate current-vs.-grid-bias voltage it will be found that there are two types of pentode vacuum tubes: *remote cutoff* and *sharp cutoff*. The respective $I_p V_b$ curves for these two varieties are shown in Fig. 3-19. The sharp-cutoff pentode has a plate-current characteristic that drops sharply to zero at some specific and relatively low value of negative grid bias. The other type of pentode is the remote-cutoff type, which is also sometimes called a *variable-mu* pentode. Notice that its plate current does not drop sharply to zero, but approaches zero in a more gradual manner. In general, it takes a very high negative bias voltage to force the plate current to zero in a remote-cutoff pentode.

The variable-mu characteristic is achieved by using the specially wound control-grid structure shown in Fig. 3-20. According to the Navy's training manual, *Basic Electronics*, Vol. 1,

> If, in a pentode, the spacing between grid wires is not uniform along the entire length of the control grid structure, the various portions of the grid will possess different degrees of electrostatic control over the plate current [See Fig. 3-20]. The remote cutoff action is due to the structure of the grid, which provides a variation in amplification factor with a change in bias. The control grid is wound with wide spacing at the center and with close spacing at the top and bottom. When weak signals and low grid bias are applied to the tube, the effect of the nonuniform turn spacing of the grid on the cathode emission and tube characteristics is essentially the same as for uniform spacing. As the grid bias is made more negative to handle

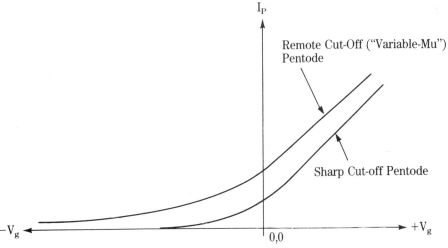

3-19 Comparison of remote and sharp cutoff pentode curves.

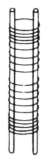

3-20 Specially wound grid for a variable-mu tube.

larger input signals, the electron flow from the sections of the cathode enclosed by the ends of the grid is cut off. The plate current and other tube characteristics are then dependent on the electron flow through the open section of the grid. This action reduces the gain of the tube so that large signals may be handled with minimum distortion.

Beam Power Tubes

According to one way of thinking, it is proper to call any large vacuum tube a *power tube*. This is often justifiable because such tubes are quite able to handle large currents at high voltages; hence they are high power. There is, however, a distinct class of power tubes used in radio transmitters and audio power amplifiers, called *beam power tubes* after their internal structure (Fig. 3-21).

Before we deal with the beam power-tube construction, let us first consider some problems of signal power generation. The main limiting factor is anode heat. No vacuum tube, or any other device for that matter, is perfect—that is, nothing is 100 percent efficient. The difference between the dc power consumed from the power supply and the signal power delivered to an external load is given up as heat

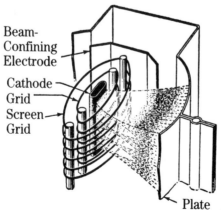

Beam-
Confining
Electrode

Cathode
Grid
Screen
Grid

Plate

3-21 Cutaway view of the beam power tube.

in the plate structure of the tube. This is partially what causes the tube's glass envelope to heat up to an unbearable temperature after only a few moments of operation. The rest of the heat comes from the filament current.

The plate dissipation rating of a vacuum tube is the amount of heat, in watts, that the tube can dissipate in normal operation without damage. It must, however, be pointed out that this means *continuous commercial service* (CCS). Tubes are frequently operated in excess of their plate dissipation ratings for short periods, *intermittent commercial and amateur service* (ICAS), even though this might mean shorter reliable life expectancy.

Most modern radio transmitters that use the tetrode type power tube are actually using *beam power tubes*. Likewise, modern audio power amplifier tubes are of the beam-power-tube construction. An example of beam-power-tube-construction is shown in Fig. 3-21. This type of power tube uses a set of *electron beam–forming plates*, or *beam deflection plates*, to bunch the electrons together and more effectively direct them towards the plate.

The beam power tube operates at better efficiency and higher power gain than other types of power tubes. They also possess a better third-order distortion figure. The low-power, but very popular, type 6146B is an example of a beam power tube—and a tetrode at that.

Other Matters to Consider

In this section we shall discuss the definition of *amplification factor* (used on an intuitive level so far), *plate resistance*, *plate impedance*, and *transconductance*.

An amplification factor is the ability of the tube to amplify small signals. It is usually denoted by the lowercase Greek letter μ (mu). The amplification factor, or u, is given by

$$\mu = \frac{\Delta V_b}{\Delta V_c} \quad (I_p = \text{constant})$$

The Greek letter Δ (delta) is used to represent a concept that you will see often in electronics, in lieu of the more correct notation from calculus. It is used to denote a "small change in . . ." It should be noted that the change should be very small, and that the formula is not totally correct unless the change approaches, but is not quite equal to zero. In that more nearly ideal case we would use calculus notation and write:

$$\mu = \frac{dV_b}{dV_c} \quad (I_p = \text{constant})$$

It is also important that plate current be held constant when this measurement is made. The other terms in the equation are the familiar plate voltage V_b and negative grid bias V_c.

It is possible to use the amplification factor to compute the approximate cutoff voltage for a vacuum tube at any given plate voltage. The equation is:

$$-V_{co} = \frac{V_b}{\mu}$$

Where: V_{co} = grid voltage necessary to effect cutoff
V_b = plate voltage
μ = amplification factor

Plate resistance and *plate impedance* are related terms, but are different in that one is a quasistatic dc process and the other is an ac process. Many who are less discerning than they should be are often tempted to look upon these as the same. Plate resistance is defined as:

$$R_p = \frac{V_b}{I_p}$$

Where: R_p = plate resistance
V_b = plate voltage (dc)
I_p = plate current (dc)

Since direct current is specified for the measurement of plate resistance, it is the static opposition to the flow of current from the cathode to the plate. Plate impedance, on the other hand, is defined in terms of alternating current, so it is a dynamic measure of the same thing.

It must be noted, however, that plate resistance varies a great deal with the plate voltage level, but the plate impedance is much more constant, especially in the all-important linear region of the characteristic curve. The formula for the plate impedance is:

$$r_p = \frac{\Delta v_b}{\Delta i_p} \; (V_c = \text{constant})$$

Note the use of lowercase letters in the plate impedance formula. This is a general convention to denote an ac case of something that could be either ac or dc.

Transconductance, or *mutual conductance* as it is often called, is a superior method for measuring vacuum tube performance. It is defined as a change in plate current for a given change in grid voltage:

$$g_m = \frac{\Delta i_p}{\Delta v_c} \bigg| V_b = \text{constant}$$

The unit of transconductance g_m is the *mho*, or more conveniently in real cases, the *micromho* (μmho). In modern texts on electronics you will encounter this same concept (conductance), but the unit of measure is the *Seimans* (S). But never fear, they didn't change anything; 1 S = 1 Mho.

One expression relates the amplification factor, plate impedance, and transconductance:

$$\mu = g_m r_p$$

One last fact that should be mentioned is that all current in both plate and the screen grid of a vacuum tube also must flow in the cathode. In other words,

$$I_{\text{cath}} = I_{\text{plate}} + I_{\text{screen}}$$

This equation may seem to be useless at this point, but you are advised to remember it because it is quite useful at times. For instance, you might know the plate current, plus the cathode resistance and the voltage across that resistance, and you may need to find the screen current.

Multipurpose Tubes

Various multipurpose tubes have been designed and offered to radio manufacturers. Figure 3-22 shows some of the more common forms of these special tubes. The purpose of the multifunction tube might be to limit the number of tubes needed in a radio.

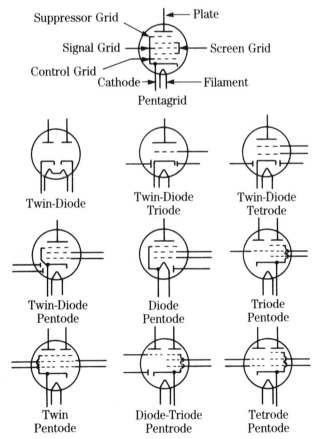

3-22   Schematic symbols of multifunction tubes.

For example, the twin diode and pentode is used to provide AM detector, AVC rectifier, and first audio preamplifier, all in one envelope. That design eliminated one tube (6AL5, 6H6, etc.) that was used in earlier designs. Certain other multipurpose tubes were designed because they allowed a different type of circuit that was a bit better than the previous arrangement. For example, the pentagrid converter replaces the local oscillator and mixer stages used in earlier radios.

Tube Bases and Envelopes

Over the years that radio tubes were being developed, several different forms of base were used. The earlier tubes were only diodes and triodes, and had directly heated cathodes, so a four-prong (or "pin") base was sufficient (two for the cathode, one for the grid, one for the anode). But as the number of grids rose, so did the number of pins required to make the tube work. Figure 3-23 shows the forms of vacuum tube base used over the years.

Pin numbering depends on the base key used on each form of base. On the older four- and six-pin tubes, two of the pins were larger than the others. With the large pins held towards you, pin No. 1 is to the left, and the numbering is clockwise. Therefore, pin No. 1 and pin No. 4 (on four-pin) or pin No. 1 and pin No. 6 (on six-pin), are the pins closest to you. The keyway on an eight-pin tube is located between pins 1 and 8. With the keyway toward you, pin 1 is on the left and pin 8 is on the right. On glass miniature tubes, the "keyway" is the space in the pin circle. Pin 1 is on the left, and either pin 7 or pin 9 is on the right.

Through the years a number of different envelope shapes were used in radios. Figure 3-24 shows a number of the most popular forms of base and envelope used.

Testing Vacuum Tubes

Variations in the grid voltage of a vacuum tube will produce changes in the plate current, i.e., cathode-to-plate electron flow. Ohm's law tells us that these changes

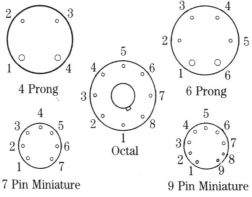

3-23 Base diagrams for various size tubes.

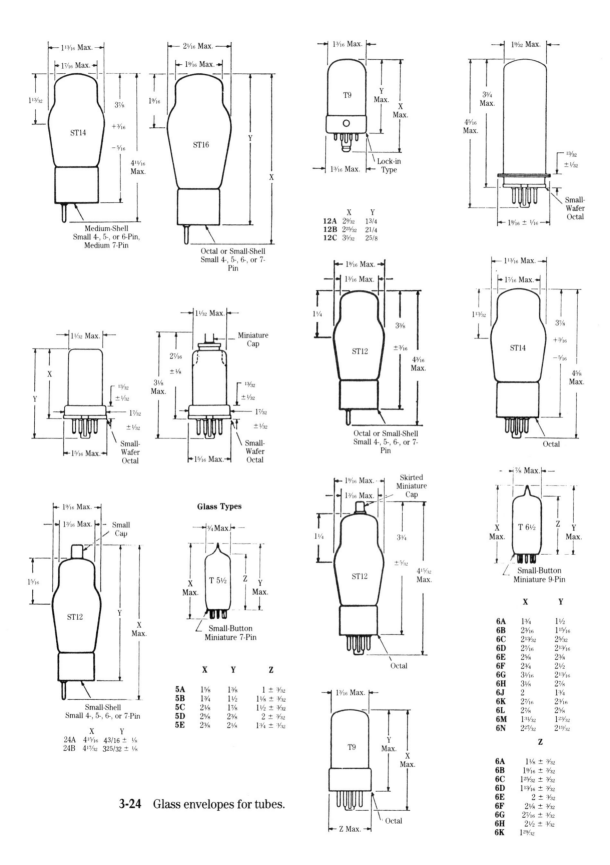

3-24 Glass envelopes for tubes.

will be reflected as changes in the voltage drop across load R_L. These variations are the source of the voltage amplification offered by the vacuum tube.

Several vacuum tube parameters are of interest when you are trying to ascertain the quality of the device. The most important testing parameters are amplification factor (called mu), plate resistance (r_p), and the mutual conductance, also called transductance (g_m).

Vacuum tube testers

Vacuum tube testers are available in several configurations, but the basic types are *short-circuit tester, emission tester*, and *transconductance tester*. These different types of instrument are capable of different levels of usefulness, but all are found in service. Some instruments favored for many years in the radio-TV service industry will test all three ways.

Short-circuit tester

Figure 3-25 shows the basic circuit for a simple short-circuit tester. This type of tester gives only very limited information, but it is useful when large numbers of tubes are being screened, and when the suspected failure mode is an interelement short circuit. This type of test circuit is almost invariably built into the same instrument as emission and transconductance testers, because those types are often not sensitive to high-resistance interelement shorts.

The short-circuit tester consists of a series of continuity testers arranged to locate any resistance paths between adjacent tube elements. Each continuity tester uses a low-current filament transformer—less than 1 ampere—and a compatible indicator lamp. In some models there is only one lamp. A switch is used to connect

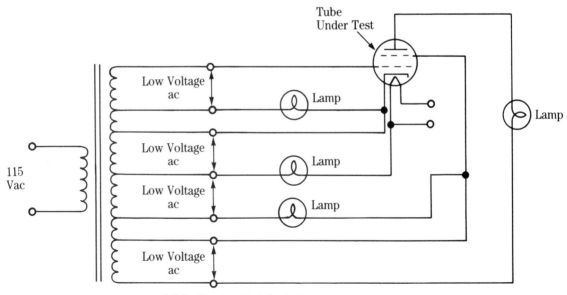

3-25   Shorts-only tube tester.

the lamp to the various elements. In those testers, the operator must switch through all positions in order to test all elements.

Caution—It is wise to wait several minutes before testing a tube for short circuits. Some shorts do not show up until the tube has reached its normal operating temperature. Unlike transistors, vacuum tubes may require two to three minutes to reach that temperature. This is also the reason why the ohmmeter will not suffice; it cannot be operated in a live circuit, so it will give a large number of false negative results; that is, it will say there are no shorts, when in fact there are. You might also want to lightly tap the tube under test, for example, with a pencil eraser end, in order to reveal intermittent shorts.

Emission testers

The simplest class of tube tester that yields a qualitative insight into the performance of any particular tube is the emission tester, an example of which is shown in Fig. 3-26. This circuit tests the tube for the emission of electrons from the cathode. The tube is connected into a diode configuration in which all elements except the filaments and cathode are tied to the plate. One principal cause of tube failure is the declining emission of electrons from the cathode. It is this phenomenon that the emission tester is designed to detect.

The tube in a circuit such as Fig. 3-26 acts very much like a diode. The emission current of the tube is defined as the saturation current of the tube, connected into the diode configuration of Fig. 3-26. If the plate voltage is increased from near zero, we find that the plate current will also increase in a very nearly linear manner, except for very low potentials.

The current will continue to increase with increases in plate voltage until a certain critical potential is reached. But past that point we find that the anode is unable to attract any further electrons; that is, all of the electrons boiled off the cathode by thermionic emission are attracted to the anode. Since any further increase in plate voltage produces little or no increase in plate current, the tube is said to be *saturated*. The current level at which this occurs is called the *saturation current* or *emission current*.

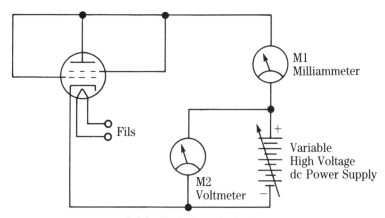

3-26   Emission tube tester.

The emission-type tube tester checks for this current, and if the actual emission current is considerably lower than it is supposed to be, indicates *reject*. Some tubes cannot be tested at the actual saturation current, either because it may be inconvenient to do so, or because current level is too high and may actually damage the tube.

These tubes are tested at some specific plate potential, at which a given current may be expected normally. If the tube will not produce at least a certain predetermined percentage of the current, such as 70 to 80 percent, then it is rejected. The emission tube-tester can spot a grossly bad tube, but it is not the most sensitive of tests. Note that most drug store testers, which were once very popular in the United States, were of the emission type because they are so very easy to operate.

Emission testers tend to create a lot of false negatives. That is, a tube is called *good* when in fact it is actually bad. It was quite common in the 1950s and 1960s, when drug store tube testers were the retail rage, to have a customer become quite irate when a defective tube showed up after he'd tested them "good" at the drug store. The problem was the limitations of the emission tester at the drug store, compared with the more effective transconductance tester at the service shop.

Transconductance testers

Defects that will prevent the tube from performing satisfactorily will not always show up on the emission tester. Therefore, the type of tube tester preferred by most professional service people is the *mutual conductance* or *transconductance tester* of Fig. 3-27. The circuit shown in Fig. 3-27A is a static transconductance tester using the *grid shift*, or *grid-level shift*, method.

Switch S1 is in position 1 at the beginning of the test, making grid voltage V_c equal to V_2 alone. Both V_1 and V_2 are adjusted to produce a convenient and safe current in the plate circuit. It is usually deemed wiser to begin with the plate voltage V_1 at some low level, well within the range of voltage that can be tolerated by the tube being tested, and keep the grid voltage considerably beyond cutoff for that tube. The operator may then adjust these voltages gradually until the plate current is at a convenient level. When this adjustment is completed, note the values of V_b, I_p, and V_c.

To perform a test, place switch S1 in position 2. This operation makes grid voltage V_c equal to $V_2 + V_3$. The plate voltage is then measured and the supply is readjusted if a change has been noted. The plate voltage at this point must be equal to the plate voltage that existed originally.

Now, read the plate current I_p and note the difference between this reading and the initial value. The transconductance is found from

$$g_m = \frac{\Delta I_p}{1.5}$$

Where: I_p = the plate current in amperes (A)
g_m = the transconductance in mhos
Δ = "a small change in . . ."

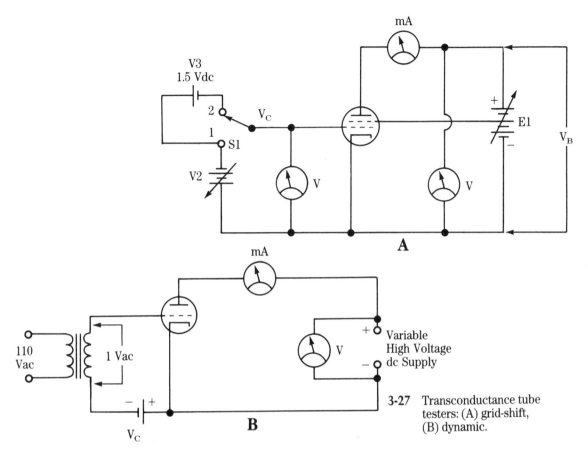

3-27 Transconductance tube testers: (A) grid-shift, (B) dynamic.

The factor 1.5 is the change in grid voltage produced by the circuit shown. If another value of ΔV_c is used, then that factor should be substituted into the equation.

A dynamic transconductance tube tester is shown in Fig. 3-27B. This circuit is the basic circuit for most professional-grade tube testers. A low-voltage ac transformer is connected in series with the grid bias supply, so the transformer secondary voltage becomes the "ΔV_c" term. The actual transconductance is measured on the plate ac milliammeter, which is conveniently calibrated in microohms on professional machines.

Commercial tube testers

Commercial tube testers were available for many years, but for the antique radio buff only used testers are easily available today. You have several options for finding a used tester. First, call older service shops that have been in business at least 15 years and ask if they have a used tube tester—an "oldie but goodie"—stuffed under a bench somewhere. Second, go to amateur radio events such as "hamfests." These (usually outdoor) events usually feature a "tailgating" section, where people sell their old radio junk, and tube testers frequently show up. Finally, shop the antique radio clubs, newsletters, and *Antique Radio Classified*. You might also place a paid classified advertisement in *Electronic Servicing & Technology* magazine.

4
Radio Receivers: The Basics

IN THE PAST THREE CHAPTERS WE DISCUSSED THE HISTORY OF RADIO RECEIVERS, THE basics of vacuum tube theory (the root of antique and classical radio technology), and the elements of radio reception. Now let's get down to the basic theories of radio receivers. We looked at this material briefly in chapter one, but will now expand the material to give you insight on what goes on inside the box. To do this job we will look at the tuned radio frequency (TRF) and superheterodyne types of receiver. First we will take a brief look at the elements of radio signals.

Radio Signals

One of the things that fascinated people about radio was the fact that it arrived seemingly by magic through the air. It was very quickly found out that radio signals are *electromagnetic waves* exactly like visible and infrared light except for frequency and wavelength. Radio waves have a much lower frequency than visible light, and therefore have wavelengths that are much longer. The wavelengths of radio signals range from around 25,000 meters at the sub-VLF range, down to millimeters in the upper microwave spectrum.

We will consider three different forms of signal: *continuous wave* (CW), *amplitude modulated* (AM), and *frequency modulated* (FM). The CW signal will only be considered very briefly because it is largely irrelevant to most readers of this book. The CW signal (Fig. 4-1) is made up of sinusoidal oscillations at the transmitter frequency.

For example, a 500 KHz maritime CW signal will oscillate 500,000 times per second. The critical thing to note about the CW signal is that it has a *constant amplitude*. If the signal is turned on and off to form dots and dashes of the Morse code, then it is possible to signal with the CW signal. This was the type of signal

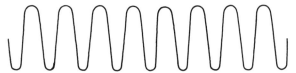

4-1 Continuous wave sinusoidal signal.

transmitted by ships at sea during the early days of wireless, and maritime radiotelegraph CW is still heard on the air today.

Modulation is the act of adding information to an unmodulated radio signal called the *carrier*. That unmodulated signal is, by the way, the same as an unkeyed CW signal. The modulating signal is usually audio from speech or music sources. In radio, the technical term for the modulating signal is "intelligence," although if one listens to certain talk shows, the Citizen's Band and other transmission of radiotelephone signals, it is difficult to keep a straight face when talking about the modulating signal as "intelligence." Figure 4-2 shows the relationships among the modulating signal (Fig. 4-2A), the carrier (Fig. 4-2B), and the resultant amplitude modulated signal (Fig. 4-2C).

The transmitter modulator stage superimposes the audio signal onto the carrier, resulting in the characteristic signal shown in Fig. 4-2C. It is this signal that is received at the radio set. Note that the peaks of the AM signal vary in step with the audio modulating signal. The frequency of the AM carrier remains constant, but its intensity, or *amplitude*, varies with the audio signal.

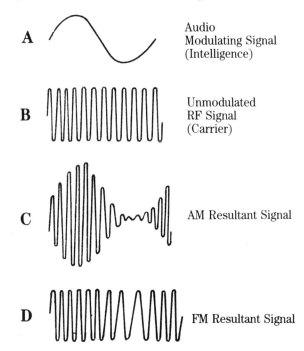

4-2 Types of signal in radio: (A) audio modulating signal, (B) unmodulated CW/carrier signal, (C) amplitude-modulated signal, (D) frequency-modulated signal.

By way of contrast, the FM signal is shown in Fig. 4-2D. In this case, the frequency, or *phase*, of the carrier varies in step with the audio signal, while the amplitude remains constant.

The Tuner

The air surrounding a radio is filled with a cacocphony of electromagnetic signals. Very early in the radio business it was found that radios not only worked better with tuning units, but would separate signals on different wavelengths (frequencies).

The standard radio tuner consists of an inductor or coil, and a capacitor or condensor. Figure 4-3A shows a tuner, while the appropriate schematic is shown in Fig. 4-3B. An inductor stores energy in a surrounding magnetic field, while the capacitor stores energy in an electrostatic field between the two plates.

Inductors and capacitors both provide some opposition to the flow of ac current, including radio frequency or RF frequencies. This opposition is called *reactance*, and for inductors it is *inductive reactance* (X_L) and for capacitors it is called *capacitive reactance* (X_c). At a certain frequency, called *resonance*, these two reactances are equal: $X_L = X_c$. When this occurs, the LC "tank" circuit can oscillate so that energy is swapped back and forth between the electrostatic field of the

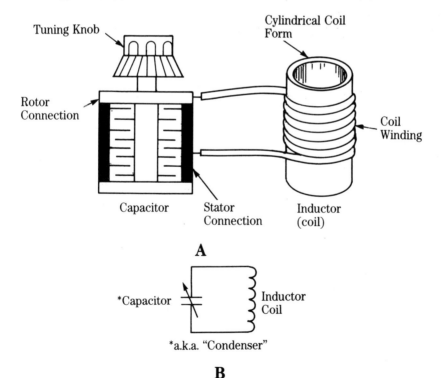

4-3 Resonant tuning circuit consists of an inductor coil and a variable capacitor: (A) physical form, (B) circuit symbols.

capacitor and the magnetic field of the inductor. It is this frequency that the tuner will accept, while rejecting others. The resonant frequency of the LC "tank" circuit of Fig. 4-3 is given by:

$$f = \frac{1}{2\pi LC}$$

Where: f = frequency in hertz (Hz)
L = inductance in henrys (H)
C = capacitance in farads (F)

Radio tuners don't normally use the simple LC resonant tank circuit of Fig. 4-3, but rather depend on the RF transformer principle shown in Fig. 4-4A and (schematically) in Fig. 4-4B. The primary winding of the RF transformer (L1A) consists of a few turns of wire wound on the lower end of the same coil form as the main tuning coil, L1B.

If either the inductor or capacitor is made variable, then an operator can select from a number of resonant frequencies by varying the frequency of the tuned circuit. In most radios the tuner control is a variable capacitor because of the ease with which these components can be made.

Figures 4-4C and 4-4D show how the circuit works. When the radio wave is captured by the antenna, a small current (I_1) at the same frequency as the radio wave is set up in the antenna-ground path. Only one-half of the radio wave causes the current to flow down towards the ground as shown in Fig. 4-4C. Because the primary (L1A) and secondary (L1B) are magnetically coupled, current I_1 induces current I_2, which flows in the LC tank circuit, producing a signal voltage (V) across the output terminals. Likewise, when the radio wave reverses polarity, the antenna-ground current also reverses direction (Fig. 4-4D). The current in the secondary is also therefore reversed, creating the other half of the output signal voltage (V). As the radio wave oscillates between its peaks at the radio transmitter frequency, the currents (I_1 and I_2) and signal voltage (V) also oscillate at the same frequency. Only the resonant frequency oscillates efficiently in the LC tank circuit, so only it will appear at the output terminals, even though many waves set up currents in the antenna-ground circuit.

Even though most radio tuners use a fixed inductor and a variable capacitor, which is ganged to the tuning knob on the radio front panel, there are a few models out there that used a fixed capacitor and a variable inductor (Fig. 4-5).

The coil has a hollow coil form, and is tuned with a powdered iron or ferrite core connected to the tuning dial through a dial cord. These radios were produced during World War II when the metal needed to make a variable capacitor was far too precious to the war effort for use in consumer radios. Very few consumer radios were made, but those that were often used the variable-inductor scheme of Fig. 4-5.

The *selectivity* of the radio is its ability to separate stations adjacent to each other on the dial. The number of LC tuned circuits in the signal path determines the selectivity. Figure 4-6 shows the curves to expect with one tuned circuit (Fig. 4-6A), two tuned circuits (Fig. 4-6B), and three tuned circuits (Fig. 4-6C). Note that in each

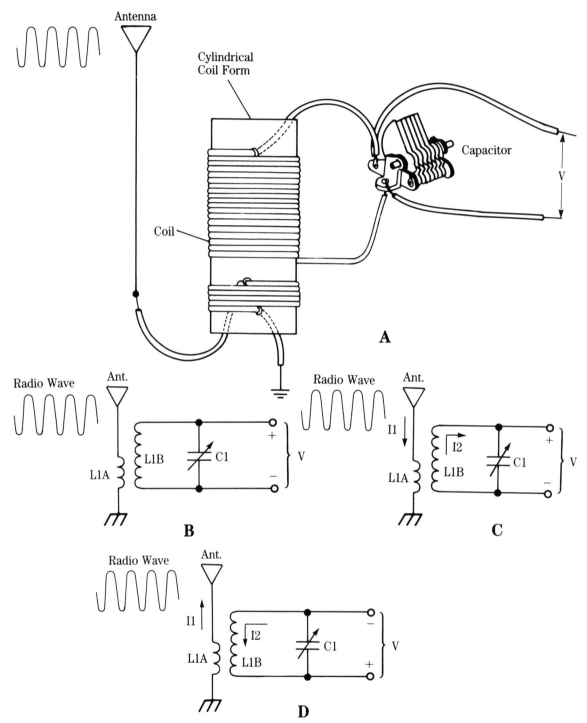

4-4   Antenna coupler tuned circuit: (A) physical form, (B) circuit symbols, (C,D) currents flowing in the antenna circuit.

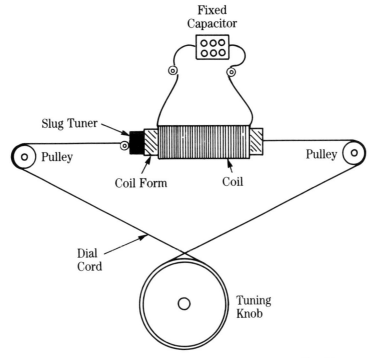

4-5 Permeability tuning mechanism popular in World War II era.

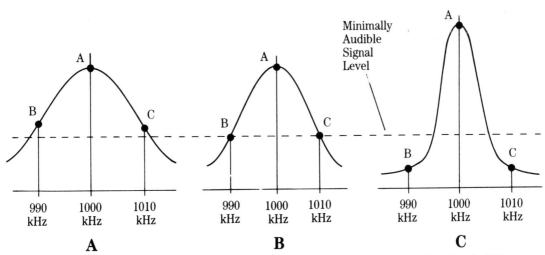

4-6 Bandpass characteristics of three cases: (A) wide bandwidth, (B) medium bandwidth, (C) narrow bandwidth.

case the LC tuned circuit does not pick out only one single frequency, but rather it picks up a small band of frequencies surrounding the resonant frequency.

There are three cases shown in Fig. 4-6. In each case the level of signal required for minimum audibility is the same, shown by the dotted line. Also, in each case there are three signals: A is at 1000 KHz, B is at 990 KHz, and C is at 1010 KHz.

In Fig. 4-6A the tuning is broad, so both interfering signals (B and C) are above the minimum audibility level. At Fig. 4-6B the interfering signals are exactly at the level of audibility, while at Fig. 4-6C the offending signals are below the level of audibility; so they will not be heard.

The selectivity of the receiver is measured in terms of the bandwidth of the tuned circuit. The standard method for specifying is to measure the frequency response of the circuit to find the points where the power drops -3 dB, or the voltage drops -6 dB (same point, but specified differently). In Fig. 4-7 we see the response of a moderately selective LC tank circuit. The bandwidth is the frequency difference F2–F1, where F1 and F2 are the frequencies at which the response drops -6 dB (voltage) from the level found at resonant frequency, F_R.

Tuned Radio-Frequency Radios

The *tuned radio-frequency* (TRF) radio uses one or more radio frequency (RF) amplifiers to boost the weak signal received from the antenna. The simplest form of TRF radio (Fig. 4-8) consists of a single RF amplifier stage, with frequency selection (LC tank) circuits at the input and output. The signal at the antenna input of the RF amplifier is very weak, but because of the amplification of the tube in the amplifier, it is boosted to a stronger level at the output.

The RF amplifier is followed by a detector, which is in turn followed by an audio amplifier and a reproducer, such as a loudspeaker. The detector demodulates the RF signal to recover the audio signal that was impressed on the RF carrier at the transmitter. This audio signal is boosted to a higher power level in the audio frequency amplifier, and then fed to a loudspeaker, earphones, or other audio reproducer.

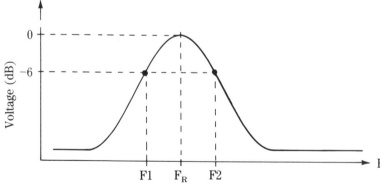

4-7   Frequency response characteristic.

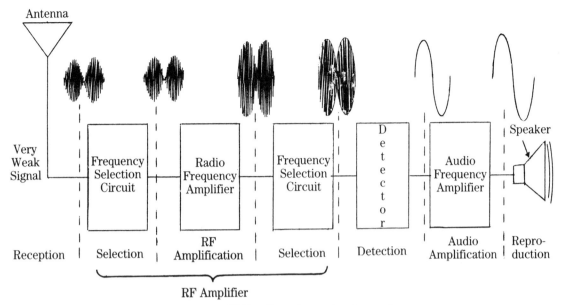

4-8   Single-stage TRF radio and associated waveforms.

More complex TRF radios will use two or more RF amplifiers cascaded (Fig. 4-9), prior to the detector and AF amplifier. The extra amplification boosts the signal level even higher than is possible in the single stage version.

Also, the additional tuned circuits tend to sharpen the selectivity. If a single tuning shaft operates all of the variable capacitors, that is, the capacitors are *ganged* together, as shown by the dotted line in Fig. 4-9, then "single-knob" operation is possible.

Superheterodyne Radios

The superheterodyne radio receiver (Fig. 4-10) was invented in the early 1920s, but originally only a very few sources could supply them because of patent restrictions. Later, however, patents were pooled as the radio industry grew up, and eventually the patents expired. The superheterodyne design was so superior that within a decade it took over all but a very few radios, and is still today the basic design of all AM and FM radio receivers.

The block diagram to a superheterodyne receiver is shown in Fig. 4-10. The basic idea of the superheterodyne is to convert the RF carrier signal from the radio wave to a lower frequency where it can be amplified and otherwise processed. The stages of the basic superheterodyne receiver consists of a *mixer*, a *local oscillator* (LO), an *intermediate frequency* (IF) *amplifier*, a *detector*, and an *AF amplifier*. The latter two stages are also found in TRF radios, and serve exactly the same function in superheterodynes. Better quality superhet radios also include an *RF amplifier*.

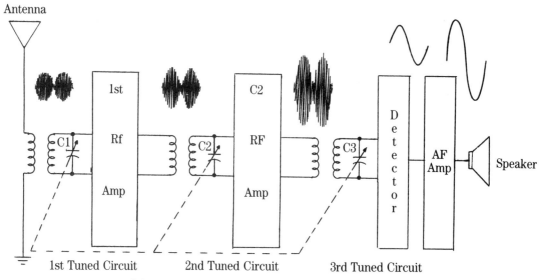

4-9   Multistage TRF radio and associated waveforms.

The RF amplifier boosts the weak signal from the antenna, and provides the radio with some additional selectivity. It also tends to prevent the signal from the LO from being coupled to the antenna, where it could be radiated into space. The RF amplifier is sometimes called the *preselector* in radio schematics.

The output of the RF amplifier is applied to the input of the mixer stage. The local oscillator (LO) signal is also input to the mixer. The two are mixed together in

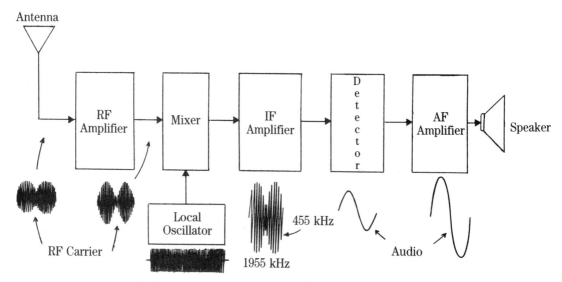

4-10   Superheterodyne radio and associated waveforms.

a nonlinear circuit, two produce at least four output frequencies: RF, LO, RF-LO (difference), and RF+LO (sum). Either the sum or difference frequency can be selected for the intermediate frequency (IF), but in the vast majority of antique or classic radios it was the difference frequency that was used. For AM radios the IF frequency is 455 KHz or 460 KHz in home radios, and 260 KHz or 262.5 KHz in audio radios. In many radios the LO and mixer are combined into a single stage called a *converter*. Another name for the mixer or converter is *first detector*.

The IF amplifier provides the radio with the largest amount of signal gain, and the tightest selectivity. It is the fact that the IF operates on only one frequency that permits very high gain to be achieved without oscillation and other difficulties.

The detector operates at the IF frequency, and will demodulate the IF frequency to recover the audio. From the detector it is passed on to the AF amplifier and the reproducer.

Radio Example

Figure 4-11 shows an RCA "All-American-Five" superheterodyne radio receiver. The description of this radio given below is quoted from *The RCA Receiving Tube Manual* from which this example is adapted.

This basic five-tube superheterodyne radio receiver operates directly from an ac power line of 117 volts or a dc supply. AC power inputs are converted to dc power by the 35W4 half-wave rectifier circuit. The receiver uses a series heater arrangement. With ON-OFF switch S1 closed, the heater string is connected directly across the 117-volt input terminals. A 6.3-volt panel lamp I1 connected between heater pins 3 and 6 of the 35W4 rectifier tube lights to indicate that power is applied to the receiver.

A ferrite-rod or loop antenna L and tuning capacitor C1 select amplitude modulated RF signals from the desired broadcast band (550 to 1600 KHz) radio station and couple these signals to grid No. 3 (pin 7) of the 12BE6 converter. A local-oscillator signal, developed by the resonant circuit formed by oscillator coil T1 and variable capacitor C5 and C6, is also applied to the 12BE6 pentagrid converter, at grid No. 1 (pin 1). The modulated-RF and local-oscillator signals are mixed across the nonlinear impedance of the converter tube to produce the 455 KHz intermediate frequency used in the receiver. The antenna and oscillator tuning capacitors C1 and C5 are mechanically ganged so that the antenna and oscillator resonant circuits can be adjusted together to maintain the 455 KHz difference frequency for any dial setting in the broadcast frequency band. Positive feedback to sustain oscillations is inductively coupled by T1 from the cathode of the 12BE6 converter to the local-oscillator resonant circuit.

A single IF stage, which uses a high-transconductance 12BA6 remote-cutoff pentode, provides the required amplification of the intermediate frequency signals. This stage is made selective at 455 KHz by the double-tuned input and output transformers T2 and T3. Audio-signal components are extracted from the IF signal by the second detector circuit, which consists of pin 6 diode section in the 12AV6 tuner and associated components. (The pin 5 diode section of the tube is not used and is shorted to the tube cathode, pin 2.) The audio output from the detector is developed across the Volume Control potentiometer, R6, which provides the sound level of the receiver. The detector also develops a negative dc voltage proportional

29-1   AC/DC SUPERHETERODYNE RADIO RECEIVER

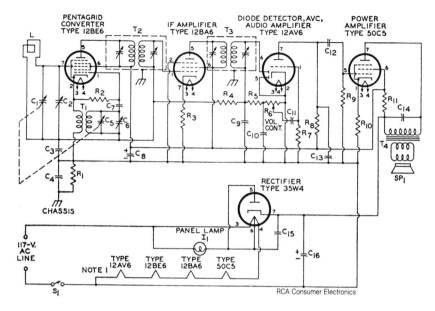

RCA Consumer Electronics

Parts List

C_1, C_5=Ganged tuning capacitors ; C_1, 10-365 pF, C_5, 7-115 pF
C_2=Trimmer capacitor, 4-30 pF
C_3=0.05 μF, paper, 50 V
C_4=0.1 μF, paper, 400 V
C_6=Trimmer capacitor, 2-17 pF
C_7=56 pF, ceramic
C_8=30 μF, electrolytic, 150 V
C_9, C_{10}=150 pF, ceramic
C_{11}, C_{14}=0.02 μF, paper, 400 V
C_{12}=0.002 μF, paper, 400 V
C_{13}=330 pF, mica

C_{15}=0.05 μF, paper, 400 V
C_{16}=50 μF, electrolytic, 150 V
I_1=Panel lamp, No. 40 or 47
L=Loop antenna or ferrite-rod antenna, 540-1600 kHz (with specified values of capacitance for C_1 and C_2)
R_1=0.22 megohm, 0.5 watt
R_2=33000 ohms, 0.5 watt
R_3=100 ohms, 0.5 watt
R_4=3.3 megohms, 0.5 watt
R_5=47000 ohms, 0.5 watt
R_6=Volume control, potentiometer, 0.5 megohm
R_7=4.7 megohms, 0.5 watt
R_8, R_9=0.47 megohm, 0.5 watt

R_{10}=150 ohms, 0.5 watt
R_{11}=1200 ohms, 1 watt
S_1=On-off switch ; single-pole, single-throw
SP_1=Speaker
T_1=Oscillator coil for use with 7-115 pF tuning capacitor and 455-kHz intermediate-frequency transformer
T_2, T_3=Intermediate-frequency transformers, 455 kHz (permeability-tuned type may be used)
T_4=Output transformer for matching impedance of voice coil to 2500-ohm load

4-11   "All-American-Five" superheterodyne radio receiver.

to the RF input across a 150 pF capacitor, C9, for automatic volume control in the receiver. The AVC voltage is used as bias for the converter and IF amplifier, and automatically controls the gain of these stages.

The audio-signal voltage at the wiper arm of the Volume Control potentiometer is amplified by the triode (audio voltage amplifier) section of the 12AV6, and is then used to drive the 50C5 audio-output stage. The output stage develops the audible output from the speaker. Audio-output transformer T4 matches the 2500-ohm plate-load impedance of the 50C5 to the speaker voice coil.

Figure 4-11 shows the waveforms that will be found in the radio of Fig. 4-11 when an oscilloscope is used as a signal tracer. The local oscillator signal (Fig. 4-12A) is a constant amplitude sinewave at a frequency of RF+IF. The waveform photo of Fig. 4-12A was taken at pin 1 of the 12BE6 converter tube. The

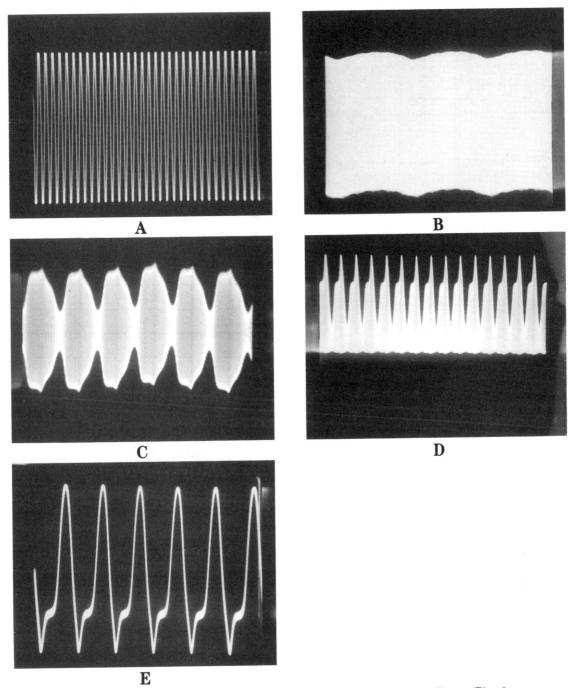

4-12 Waveforms in stages of superheterodyne receiver: (A) local oscillator, (B) mixer output, (C) IF amplifier, (D) detector, (E) detector output.

waveform at the plate of the 12BE6 is shown in Fig. 4-12B. Here we see the signal is partially modulated by the RF signal-modulation waveform. The reason why the individual RF cycles don't show in Fig. 4-12B, while they did show in Fig. 4-12A, is that the time-base of the oscilloscope was adjusted to expand the LO in Fig. 4-12A, while it was adjusted to show the 400 Hz modulation in Fig. 4-12B.

The IF waveform is shown in Fig. 4-12C, and is merely the standard amplitude modulation waveform. The irregularity of the waveform is due to problems in the signal generator, not in the radio. The signal at the plate of the detector (pin 6, 12AV6) is shown in Fig. 4-12D, while the recovered audio is shown in Fig. 4-12E. Slightly different time bases were used in order to "sync up" the oscilloscope.

5

Radio Power-Supply Circuits

THE DC POWER SUPPLY IN A RADIO RECEIVER PROVIDES THE PROPER DC VOLTAGES AND currents needed to operate the circuits. Over the years, radio receiver dc power supplies evolved from simple battery-operated designs to high-voltage dc power supplies designed to be operated from the ac power mains.

Battery-Operated Sets

In the very oldest home radios, the dc power supply consisted of three batteries, labeled A, B, and C. The A battery was a low-voltage, high-current model that was used to power the filaments of the vacuum tubes in the receiver. The B battery was a high-voltage (22.5 to 90 Vdc) battery that supplied current to the anode-cathode circuit. The C battery was a lower voltage battery that provided the tube's negative grid bias in some models. These voltages are often referred to as V_a, V_b, and V_c.

At one time, many of the public electrical power systems around the country were direct current, and were operated by the local trolley car company. They generated dc to operate their trolleys, and so took advantage of their own distribution system to serve paying customers along the right of way. People who lived in areas served by a dc power system could not use some of the nonbattery radios made after about the mid-1920s. Those people required a radio model with an ac/dc power supply. These circuits were used right up to the transistor revolution— and are still found in some designs.

Still another category of radio found in the early period was designed for small town, rural, and farm use. Until the farm belt was electrified, ac power was found only in and around cities and larger towns. If a small town or farm family wanted ac, then they had to generate it themselves, using a light plant generator; and that was costly.

A number of radios made in the 1920s and 1930s for farm use operated from 6 Vdc to 32 Vdc storage batteries. Even today there are people who can remember dad making the trip to town on Saturday morning to get the radio batteries charged in time for that evening's broadcast of "The Grand Ol' Opry" or some other vintage radio favorite.

A word of warning is necessary regarding those old farm radios. Many of them were fitted with power plugs that will plug into the 115 Vac wall outlet. *Don't plug any classic radio into 115 Vac until you are certain that it was designed for that voltage!* If the filaments of the vacuum tubes are rated for 30 volts (check tube manual), or if they are lower voltage types connected in series, and the filament voltages total 30–32 volts, but there does not seem to be any power transformer or evidence of a 115-volt ac/dc power supply, then assume that the radio is a 32 Vdc model until proven otherwise.

AC Power Supplies

The purpose of the power supply in ac operated radios is to convert the alternating current supplied to the residential power system to the various A, B, and C voltages needed to operate the radio receiver. One complete ac sinewave cycle consists of a complete positive excursion and a complete negative excursion. If it takes time T to complete one cycle, then the *frequency* is $1/T$. In the United States, the standard ac frequency is 60 cycles per second (60 Hz), while overseas 50 Hz ac is common.

The nominal ac voltage supplied to residences is 115 Vac, although this "nominal" voltage actually resides within a range of 105 volts to 125 volts. Also, *ac volts* must be defined. The standard ac waveform supplied by the power company is the sine wave (Fig. 5-1). Alternating current is *bidirectional*. The voltage rises in a positive direction to a peak value $(+V_p)$, and then falls back to zero. It then reverses direction and climbs to a negative peak, $-V_p$. Following the negative peak the ac sine waveform drops back to zero and begins to go positive again on the next cycle.

There are at least three voltage measurements typically used to describe this standard sinusoidal waveform: *peak, rms*, and *peak to peak*. The peak voltage (V_p) is measured from the 0 V baseline to either the positive or negative peak, while the peak-to-peak voltage is measured from the positive peak to the negative peak, or $+V_p - -V_p$. Thus, for the standard symmetrical sinewave, $V_{p-p} = 2V_p$.

The rms (root mean square) voltage is the voltage normally cited when talking about ac. For example, "115 Vac" really means "115 Vac rms." Although the most rigorous definition of rms voltage uses calculus, for sinewaves, V_{rms} is:

$$V_p\sqrt{2} = \frac{V_p/1.414}{0.707V_p}$$

For our purposes, then, rms ac voltage is 0.707 times the peak voltage. The rms voltage definition is derived from the amount of equivalent work that the sinewave power can do. The rms voltage is equal to the dc voltage that will perform the same amount of heating in a resistive load.

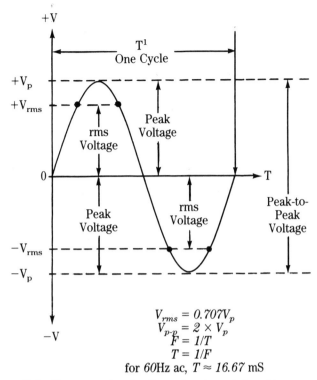

$$V_{rms} = 0.707V_p$$
$$V_{p-p} = 2 \times V_p$$
$$F = 1/T$$
$$T = 1/F$$
$$\text{for 60Hz ac, } T \approx 16.67 \text{ mS}$$

5-1 Sinewave signal voltage showing relationships.

Components of the AC Power Supply

Although there are variations from one model to another, the principal components used in ac radio power supplies are: *transformer, rectifier, ripple filter, switches, fuses,* and in some models, a *voltage regulator.* In the sections following we will examine each of these in turn.

Transformers

A transformer (Fig. 5-2A) consists of two coils of wire closely wound around a common magnetic core. One coil is called the *primary winding* and the other is called the *secondary winding.* The coils are arranged to maximize the *magnetic coupling,* i.e., the extent to which the magnetic flux surrounding the primary cuts across the coils of the secondary winding. Magnetic coupling is enhanced by winding the coils on a suitable magnetic transformer core.

The magnetic core of the transformer is made of laminated iron sheets clamped together (Fig. 5-2B). The reason for the laminated construction is to prevent *eddy current* losses from decreasing transformer efficiency.

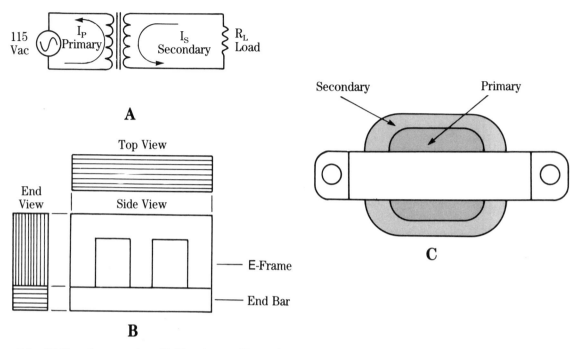

5-2   (A) Transformer action. (B) Transformer "E-core" construction. (C) Usual locations of primary and secondary.

The core of the transformer can take on any of several shapes, but one of the most common is the E-frame shown in Fig. 5-2B. The core consists of two pieces. One is the actual E-shaped frame, while the other is an end bar that will magnetically short the three open ends of the E-frame together once the windings are installed.

As mentioned previously, there are two windings on the simplest transformer: primary and secondary. The primary winding is connected to the ac source, while the secondary winding is connected to the load. In most transformers, the primary winding is wound closest to the core (Fig. 5-2C).

The primary and secondary voltages of the transformer are related by the ratio of primary to secondary turns:

$$\frac{V_{pri}}{V_{sec}} = \frac{N_{pri}}{N_{sec}}$$

Where: V_{pri} = voltage applied to the primary winding
V_{sec} = voltage appearing across the secondary winding
N_{pri} = number of turns of wire in the primary winding
N_{sec} = number of turns of wire in the secondary winding

Thus, a 6.3 Vac transformer used for powering 6 volt tube filaments from the 115 Vac power line will have a turns ratio of 115/6.3, or about 18:1. That means the primary will have 18 turns for each turn in the secondary.

If the secondary voltage is higher than the primary voltage, the transformer is called a *step-up transformer*, but if the secondary voltage is lower than the primary voltage it is a *step-down transformer*.

Many power transformers used in radio sets are actually multiple transformers and have both step-up and step-down windings. In other words, they will have more than one secondary winding (Fig. 5-3). There will be one or more high-voltage secondaries (often center tapped), one or more 6.3 Vac filament voltage secondaries to supply the filaments of the vacuum tubes in the set, and sometimes a 5 Vac winding to supply filament power to the rectifier tube used in the set.

The various windings are usually color-coded according to the scheme shown in Fig. 5-3, or some close variant. This scheme was called the *RMA transformer color code* because it was a standard of the Radio Manufacturers Association. In general: Primary Winding = Black; High-Voltage Secondary = Red; First 6.3 Vac Filament Secondary = Green; Second 6.3 Vac Filament Secondary = Brown; and 5 Vac Rectifier Filament Secondary = Yellow.

If the winding is center-tapped, when the center-tap wire will be the same color as the main wires for that winding, but will also have a yellow tracer, except on the yellow 5 Vac winding, where the tracer is some other color, usually black. The standard way of writing the colors on the center-tap wire is to list the main body color first, and then the tracer. Therefore, a "green/yellow" wire has a green body with a yellow tracer—and that particular example is a 6.3 Vac (CT) filament winding.

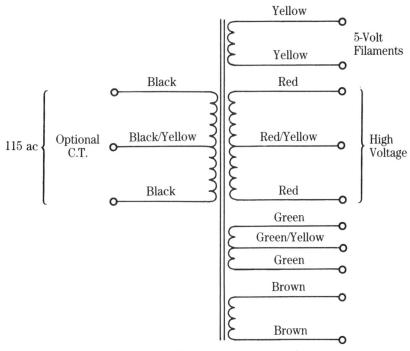

5-3   Standard transformer color code.

The voltage rating of the secondary is given as the rms value of the end-to-end voltage. If the transformer secondary is center tapped, then the secondary rating might be listed in the form "500 Vct" (volts center-tapped), or "250-0-250." Both ratings denote the same transformer, but in the latter case the voltages are not measured end to end, but rather from the center tap to each end.

The current ratio on a transformer is just the opposite of the voltage ratio. For a constant primary volt-ampere (VA) level, which is wattage if you ignore the power factor, current is stepped down in the same ratio as voltage is stepped up, and vice versa.

Transformer construction

Figure 5-4 shows several different kinds of transformer construction. Three different forms of transformer are shown here. Figure 5-4A shows an open E-frame transformer, of the sort shown earlier in Fig. 5-2C. This form of transformer is used most often for small power transformers and small filament transformers. Mounting is by way of a pair of mounting tabs on each end of the transformer. Each tab has a hole in it to accept a machine screw. The outer winding, which is the secondary, is usually covered with either wax paper or fishpaper in order to protect the enameled wire winding underneath.

The type number and/or ratings of the transformer are often printed on this covering. In most cases, the primary winding wires come out on one side of the frame, while the secondary wires are on the other side of the frame.

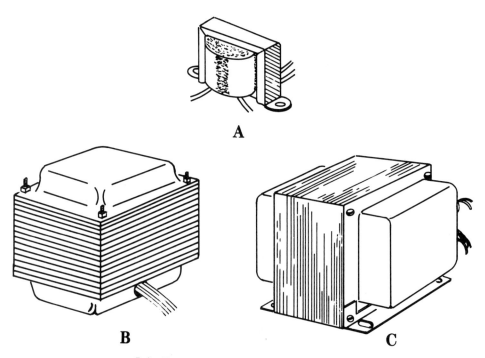

5-4   Forms of transformer construction.

Figure 5-4B shows a horizontally mounted *half-shell* transformer. This type places metal shields over the windings. While the half-shell transformer is still an E-frame design in most cases, the windings are mounted horizontally with respect to the radio chassis. Usually, a large, nearly square hole is punched in the chassis to allow one side of the shell to be inserted down into it.

The same screws that hold the shells onto the fame and hold the frame together are also used to mount the transformer to the chassis. The wires exit the transformer through a hole in one or two sides of the half-shell that faces the underside of the chassis.

The final type of transformer is the upright half-shell transformer. This type is similar to the horizontally mounted form (Fig. 5-4B), but is rotated 90 degrees. The mounting brackets are attached along one edge, and the wires are brought out through the shell on one side. The transformer pictured in Fig. 5-4C has the long axis in the horizontal plane, but there are also considerable numbers of which the long axis is in the vertical plane.

It is frequently possible to interchange transformers such as Figs. 5-4B and 5-4C by fashioning the right mountings, or deleting existing mountings. For example, you can make a horizontal half-shell by removing the mounting brackets from a model such as Fig. 5-4C. Likewise, adding a bracket to a horizontal half-shell such as Fig. 5-4B makes a dandy upright transformer (Fig. 5-4C).

Is that transformer shorted?

The power transformer in a radio set can be a little difficult to "wring out" with only an ohmmeter. In the case of a massively shorted winding, the transformer will probably grossly overheat, may smoke, ooze tar, and will probably blow the radio fuse. But what about the in-between case where a winding is only partially shorted?

Figure 5-5 shows a method for testing the radio transformer. First, either disconnect all the secondary windings, or remote all the tubes from the radio: the idea is to reduce the normal current drain. Next, connect the radio ac line cord to an outlet box in which a 25-watt to 40-watt, 115 Vac lamp is connected in series with one line of the outlet. Turn the radio on, apply ac power to the outlet box and note the

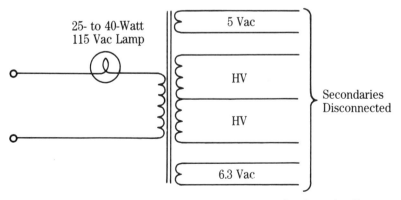

5-5 Circuit for testing an unknown transformer for short circuits.

brightness of the lamp. A good transformer will cause the lamp to barely glow, if the transformer is still connected into the circuit, while a bad transformer will cause the lamp to glow brightly.

Rectifiers

The purpose of the *rectifier* is to convert bidirectional ac to a unidirectional current. There are two generic types of rectifier circuits, called *half-wave rectifiers* and *full-wave rectifiers*. There are also three basic rectifier devices: *mechanical, vacuum tube,* and *solid-state*.

Mechanical rectifiers are no longer used, but were used industrially until the early 1950s. The only significant form of mechanical rectifier used in radios, however, were the *synchronous vibrators* used in certain antique car radios. We will therefore discuss only tube and solid-state rectifier devices.

Half-wave rectifiers   The simplest form of rectifier is the half-wave rectifier circuit. These circuits depend on a diode tube (Fig. 5-6A). Although the version shown in Fig. 5-6A uses an indirectly heated cathode, both directly heated and indirectly heated cathode rectifiers are used in common dc power supplies. In the rest of this discussion we will delete the filament connections and show only the cathode and anode.

Recall the operation of the diode:

1. When the anode is positive with respect to the cathode, current is conducted from cathode to anode; and
2. When the anode is negative with respect to the cathode, the tube is cut off and no current flows from cathode to anode.

Figure 5-6B shows the rectifier diode under the first condition. The anode is positive with respect to the cathode. In this circuit, the load is connected between the cathode and the return line to the power supply.

The polarity of the applied ac is such that the anode (plate) is positive with respect to the cathode. The tube is therefore turned on, and current flows from the negative side of the power source, through the return line, through the load, through the tube and back to the positive side of the power source. On the next half-cycle of the applied ac (Fig. 5-6C), the polarities are reversed. In this case, the anode is negative with respect to the cathode, so the tube is cut off. No current flows in the load on the negative half-cycle of the input ac waveform.

Figure 5-7 shows the relationship between the input ac waveform (top trace) and the output waveform (bottom trace). Note that there is an output only when the ac cycle is positive. The average output voltage is approximately 0.45 times the rms voltage of the applied ac waveform. The output waveform intervals where the ac waveform is negative are at zero volts. Since this is a bit wasteful of energy, the transformer used for half-wave rectifiers needs to be rated for about 40 percent more volt-amperes than transformers used for the full-wave rectifiers discussed below.

The output waveform shown in Fig. 5-7 is not the pure dc required to operate electronic circuits, but is known rather as *pulsating dc*. The frequency of the

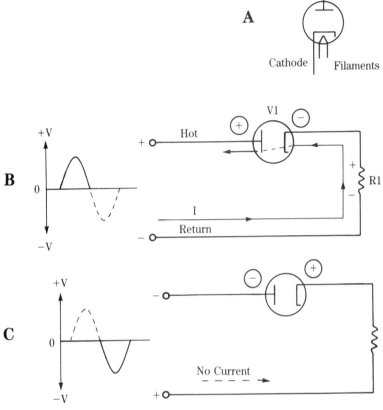

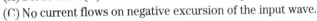

5-6 (A) Diode tube. (B) Current flow in rectifier on positive excursion of input wave. (C) No current flows on negative excursion of the input wave.

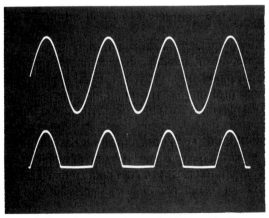

5-7 Relationship between input sinewave (upper trace) and half-wave pulsating dc output wave.

pulsating dc waveform is the same as the applied ac frequency, or 60 Hz in the United States. The departure of the pulsating dc waveform from ideal dc is measured by the *ripple factor*, of which more will be said later.

Full-wave rectifiers   The half-wave rectifier is somewhat problematic for some electronic circuits, but these problems are solved for the most part by the full-wave rectifier circuit (Fig. 5-8). The full wave rectifier uses a special power transformer that has a center-tapped secondary. The center-tap is the 0-volt reference point, so it is often grounded. Because of this, the opposite ends ("A" and "B") are at opposite polarities, even though the magnitude of the voltages is the same. Thus, when point "A" is at +150 Vac, point "B" is at −150 Vac.

Two diodes are used in the full-wave rectifier. The cathodes of these tubes (*V1* and *V2*) are connected together, while the plates are connected to opposite ends of the power transformer secondary. As a result of this connection, one anode will be positive with respect to the cathodes, while the other is negative with respect to the cathodes.

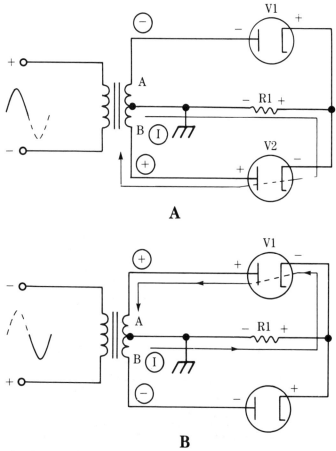

5-8   Two diode tubes in full-wave rectifier circuit: (A) current flow on positive excursion, (B) current flow on negative excursion.

Thus, one tube conducts and the other tube is turned off on each half-cycle of the applied ac waveform. On the alternate half-cycle the roles of the tubes reverse. The load, shown here as resistance R1, is connected between the common cathode connection between the tubes and the center-tap of the transformer secondary.

Consider Fig. 5-8A. In this case, point "A" is negative with respect to ground, and point "B" is positive with respect to ground. Thus, vacuum tube V1 is turned off and V2 is turned on. Current flows from the center-tap of the transformer, through load resistor R1, through the cathode-plate path of V2, and back to the transformer at point "B."

On the next half-cycle (Fig. 5-8B) the roles are reversed. Now point "A" is positive with respect to ground, and point "B" is negative with respect to ground. Thus, V1 is turned on and V2 is turned off. Current leaves the transformer at the center-tap, passes through load resistor R1, through the cathode-anode path of V1, and back to the transformer at point "A."

Now notice what happened to the current direction through load R1 on both half-cycles (compare the arrow directions in Figs. 5-8A and 5-8B). The current flows through the load in the same direction on both half-cycles of the applied ac waveform. Thus, the output waveform uses both halves of the ac cycle (see Fig. 5-9). Again, both the input ac waveform and the pulsating dc output waveform are shown. In this double-humped waveform the average voltage across the load is 0.90 times the applied rms voltage, and the ripple frequency is twice the applied ac frequency, or 120 Hz in the United States.

The ripple frequency of the full-wave rectified power supply offers a trouble-shooting approach for hum problems. If the hum is due to power-supply problems, then the hum will be at 120 Hz, but if it is due to short circuits in other tubes or coupling to the ac line, then the hum frequency is more likely to be 60 Hz.

Rectifier tube types   There are a number of different types of tubes used as both half-wave and full-wave rectifiers. In the later series of tube types, those that came along in the 1930s after the two-digit numbering system was dropped, rectifier

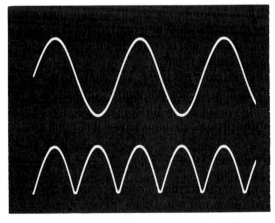

5-9   Relationship between sinewave input (upper trace) and full-wave pulsating dc output (bottom wave).

5-10   Common rectifier tubes.

type numbers tended to be from the latter half of the alphabet. This is shown in Fig. 5-10, where the type numbers are 5U4, 5Y3, 6X4, and 35W4. The rectifiers shown in Fig. 5-10 are very common in radio designs, but are not the only forms known.

Ripple filtering

The pulsating dc produced by the rectifier is not suitable for powering most electronic circuits. If this form of power is used in a radio or audio amplifier, the output will contain a large hum component. The purpose of the *ripple filter* circuit is to smooth the ripple output of the rectifier circuit. Various types of ripple filters are used, but all of them use some combination of capacitors, resistors, and in some cases inductors.

Brute force filters   The simplest form of ripple filter is the "brute-force" filter circuit of Fig. 5-11A. The filter consists of a single capacitor in parallel with the load (shown here as R_L). When the pulsating dc waveform voltage is increasing, the rectifier output current charges capacitor C1, and also supplies the load current.

The capacitor will charge to approximately V_p during this period. But after the peak has passed, the capacitor voltage is now greater than the rectifier output voltage. This means that the current stored in the capacitor will begin to flow in the load. This action is shown in Fig. 5-11B. The shaded area represents the region where energy is dumped back into the circuit, filling in the void between peaks from the rectifier.

The total load current will be due partially to the rectifier current, and partially to the capacitor current. But the contribution of the rectifier current to the total load current becomes less and less, diminishing to zero at the conclusion of the half-cycle. The process repeats itself as the next half-cycle begins.

The filter tends to smooth out the pulsations of the rectifier output, but it does not reduce it to zero. There is still a certain ripple factor present (r). The ripple factor can be defined as the ratio of the ripple component to the average dc voltage.

We can calculate the ripple factor at the output of capacitor filters from the expression:

$$r = \frac{1}{2\sqrt{3}\ FCR_L}$$

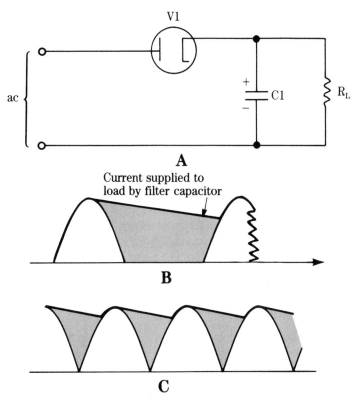

A

Current supplied to
load by filter capacitor

B

C

5-11 (A) Simple "Brute Force" ripple filter uses a single capacitor across load. (B) Ripple
filtering occurs because capacitor dumps energy into the circuit to fill in gaps between
halfwave rectified pulsating dc peaks. (C) Same case for full-wave rectified power
supplies.

Where: r = ripple factor
       F = ripple frequency—60 Hz in half-wave supplies and 120 IIz in full-wave
         supplies
       C = capacitance in microfarads
      R_L = load resistance in ohms (note: $R_L = V_{dc}/I_{out}$)

For 60 Hz ac power mains, which are common in the United States and Canada,
the equation above reduces to the following:

Half-wave Case:

$$r = \frac{1{,}000{,}000}{208 \ C_{\mu F} \ R_L}$$

Full-wave Case:

$$r = \frac{1{,}000{,}000}{208 \ C_{\mu F} \ R_L}$$

In the previous expressions, the numerator is 1 million instead of 1, so the capacitance is in microfarads, which is the unit most commonly used in radio and audio equipment. We sometimes see this basic equation in the form of percentage ripple, in which case percent ripple is merely 100*r*.

In the preceeding equation, the factor 416 for the full-wave case is twice that of the factor 208 used in the half-wave case. That means that a given size of capacitor will produce half the ripple in full-wave as it does in half-wave rectifier circuits. Conversely, the amount of capacitance needed to suppress the ripple in a full-wave supply is half that of the half-wave case. Figure 5-11C shows why this is so. Note that in this full-wave case there is considerably less space in the waveform for the filter to fill in.

Resistor-capacitor filters   In many cases, including most of the radio receivers discussed in this book, the brute force filter does not provide sufficient ripple suppression for any circuit except the audio output power amplifier. Improved ripple filtering is available from the *resistor-capacitor* (RC) filter, which will cause substantial reduction in output ripple.

Figure 5-12 shows a full-wave rectifier circuit that uses an RC ripple filter. The rectifier used in this circuit (typically 5U4 or 5Y3) does not have a separate cathode, but uses the filament as the cathode instead. The ripple filter consists of capacitor C3 and C4, plus resistor R1.

There are two output voltages available from this dc power supply. HV1 is the higher of the two, and is the potential used to power the audio power amplifier

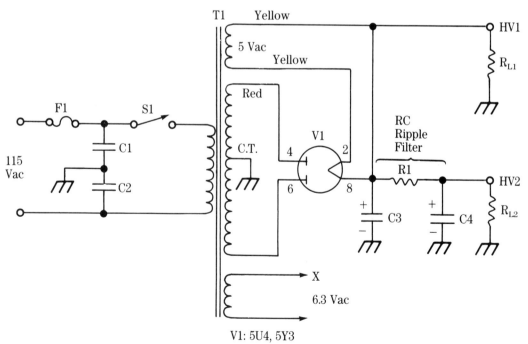

V1: 5U4, 5Y3

5-12   Typical radio dc power supply. HV1 is Brute Force filtered, while HV2 is RC filtered.

Components of the AC Power Supply **83**

stages. Potential HV2 is lower than HV1, and is used to power audio preamplifiers, RF/IF amplifiers, converters and other circuits that require a greater degree of ripple suppression, but offer relatively constant current drain. The ripple factor of HV1 is the same as for the brute force filter preceeding.

The ripple of HV2 is:

Half-wave Circuits:

$$r = \frac{1}{C_3\, C_4\, R_1\, R_L}$$

Full-wave Circuits:

$$r = \frac{2.5 \times 10^{-6}}{C_3\, C_4\, R_1\, R_L}$$

The RC filter clearly offers a much lower ripple factor, but there are also some problems. The resistor in the RC filter tends to drop the voltage quite a bit, and also severely limits the available current. Some of these problems are overcome by the *inductor-capacitor ripple filter*.

Inductor-capacitor ripple filters   Many antique radio receivers, as well as high-power radio transmitters, use inductive ripple filters in their dc power supplies. Inductive ripple filters come in several varieties. Some might be inductor-input L-sections (Fig. 5-13A), capacitor-input pi-sections (Fig. 5-13B), or multiple-section circuits (Fig. 5-13C).

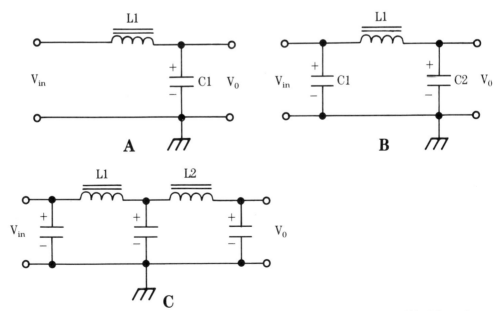

5-13   Inductor ripple filter circuits: (A) single L-section, (B) pi-section, (C) pi-L section.

Inductors respond to the rate of change of current flowing in the windings of the coil. The voltage drop across the inductor is nearly zero (there is always some small voltage drop due to the dc resistance of the winding) when the current is constant. But when the current changes, there will be a dynamic voltage across the inductor that is proportional to the inductance of the coil and the rate of change of current flowing in the coil.

The rate of change of current becomes very large close to zero crossings (when *I* is near zero). This situation results in a very high voltage drop across the inductor. At the peak of the pulsating dc, however, the rate of change is smaller, and indeed becomes zero for an instant. The voltage drop across the inductor is zero at that point.

Thus, when a varying ripple current flows in the power-supply circuit, there will be a variable voltage drop across the inductor that is in step with the ripple waveform. The current filling of the spaces between the pulsating dc pulses is from the energy stored in the magnetic field surrounding the inductor.

The ripple factors of these filters are as follows:

L-Section filter (Fig. 5–13A), full-wave rectified supplies:

$$r = \frac{8.3 \times 10^{-7}}{LC}$$

(*L* is in henrys, *C* is in farads.)

Pi-Section filter (Fig. 5–13B), full-wave rectified supplies:

$$r = \frac{3.3 \times 10^{-8}}{C1\ C2\ L1\ R_L}$$

Multiple-section filters are able to produce a much lower ripple factor than single section filters. The exact amount of ripple is a function of the types of sections used and the number of them (Fig. 5-13C is a pi-section cascaded with an L-section). For identical filter sections, the ripple falls off as the *nth* power of the ripple reduction of one section, where *n* is the number of identical sections. Other situations become more difficult to calculate.

Many older radio receivers used an inductive filter, in which the inductor used in the filter is also the electromagnet used in the loudspeaker (Fig. 5-14A). The dc current flowing in the inductor provides the magnetic field needed to operate the loudspeaker.

Figure 5-14A shows the circuit for the inductor/magnet, while Fig. 5-14B shows an example of such a loudspeaker. The inductor is the black object coaxial to the speaker cone. The other inductor-like device mounted on an attached side bracket is the audio output transformer that transforms the audio power-tube plate impedance to the lower speaker impedance.

Some radios used a second winding on the loudspeaker electromagnet called a *hum-bucking coil*. This coil magnetically picks up the ripple, which can cause a hum

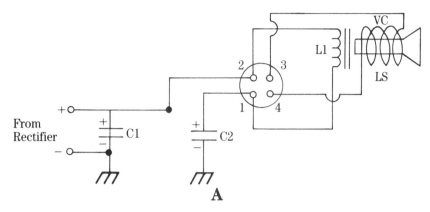

5-14 (A) Wiring of an electromagnet speaker magnet coil as a ripple filter choke. (B) Electromagnet speaker.

in the loudspeaker, and sends it to the loudspeaker voice coil. The hum-bucking coil is wired in series—opposing the voice coil in order to cancel out the hum component.

A rule of thumb used to calculate the approximate inductance needed in pi-section and L-section filters is that the inductance in henrys should be greater than or equal to $R_L/1100$.

Capacitors used in ripple-filter circuits Over the decades a number of different types of filter capacitors were used in radio receiver power supplies. In certain early radios, and in military radios, the power-supply filter capacitors were oil-filled types. These capacitors were usually rated from 2 µF to about 20 µF, and were usually in square or rectangular metal cans.

Most capacitors used in ripple filtering are *aluminum electrolytics*. In those capacitors, the "plates" are an aluminum foil covered with an aluminum oxide compound. One difference between the oil-filled capacitor and the electrolytic capacitors, the "plates" are an aluminum foil covered with an aluminum oxide compound. One difference between the oil-filled capacitor and the electrolytic capacitor is that the electrolytic is polarity sensitive. The capacitor will be marked so that the positive and negative leads can be identified. The most obvious marking shows the "+" symbol for the positive lead or terminal, and either "−," a black band, black wire, or no marking for the negative lead.

Warning! It is very important that you observe the polarity markings on electrolytic capacitors. If an electrolytic capacitor is wired into the circuit backwards it may explode and throw off fiber material and metal fragments.

There are two main ratings to observe on the electrolytic capacitor: capacitance and dc working voltage. The capacitance is usually stated in *microfarads* (μF), and has typical values from 2 μF to 250 μF. The general rule for replacement purposes is to select a capacitance that is at least as much as the original, or more.

It is generally not a problem to use more capacitance than originally called for in the radio power supply. For example, if you are trying to replace a 40 μF capacitor, there is no problem using anything larger than 40 μF, up to about two or three times the original capacitance. Larger capacitances can also be used, but one must question the stress on the rectifier during the initial "cold charge" period right after the radio is turned on.

The dc working voltage (WVdc) is the maximum sustained dc voltage that the capacitor can bear. This voltage is an absolute maximum, and must not be exceeded. In fact, there is good evidence to suggest that it should not even be approached. Because of normal manufacturing tolerances, assume that the real working voltage is about 10 percent less. For example, if you have a 250 WVdc capacitor, assume that the actual working voltage is 25 volts less, or about 225 WVdc.

Although it is probably impossible to catalog all forms the electrolytic capacitor has taken over the years, Fig. 5-15 shows some of the more popular types. An axial lead *tubular electrolytic* capacitor is shown in Fig. 5-15A. This type of capacitor may have a body diameter from 0.5 inch, to as much as 3 inches.

In most power-supply filter capacitors, however, the body was usually 1 inch in diameter. The body is an aluminum can covered with paper or some other insulating material. The positive and negative leads exit the capacitor body at opposite ends of the case. The positive end is marked "+" and the negative end is unmarked.

If the ends are not properly marked, then there is another means for identifying the positive and negative ends. The negative lead is simply spot-welded to the metal can, while the positive lead is attached to a small metal electrode mounted on a cardboard insulating disk or wafer. If you examine the ends of the capacitor, the cardboard end is the positive lead and the aluminum end is the negative lead.

Another form of tubular capacitor is the radial-lead form shown in Fig. 5-15B. This type of capacitor is found in approximately the same sizes as the axial-lead form, but has the electrical connections (either terminals or wires) coming out the same end of the body. These capacitors generally have the "+" symbol to mark the positive terminal, but in some cases it might be the negative terminal that is marked with a dark stripe down the side of the case.

Heavier radial-lead electrolytics such as Fig. 5-15C generally have larger terminals on one end. These capacitors are rare in early radios, but are found in high

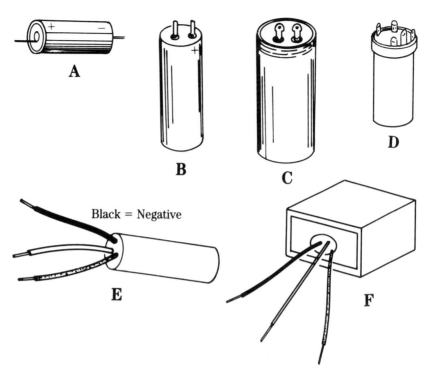

5-15 Typical filter capacitor types: (A) axial-lead tubular, (B) end-lead tubular, (C) large end-lead tubular, (D) large chassis-mounted, (E) multisection tubular, (F) older form of box capacitor.

fidelity equipment from the late 1950s and early 1960s. This type of capacitor was especially common when solid-state power amplifiers became available.

Most radio receivers did not use single-section electrolytic capacitors, but rather used multi-section models such as Fig. 5-15D. These capacitors might have two, three, or four separate capacitors in the same package, each sharing the same common negative lead. In the case of the metal can type multisection electrolytic of Fig. 5-15D the aluminum case is the negative lead.

Solder tabs are provided for the negative connection. These same tabs, incidentally, also serve to mount the capacitor to the chassis. The metal chassis will be punched out in a manner that provides slots for the mounting tabs. Alternatively, the chassis will be cut with a round hole, and either an insulated or metallic mounting adapter provided.

Another form of multisection electrolytic used in radio receivers is the tubular type shown in Fig. 5-15E. In most models, all leads, including the common negative, came out the same end. However, I recall seeing more than a few in which the positive leads came out one end, and the single common negative lead came out the other.

Still another form of capacitor found on some 1930s radios is the square electrolytic capacitor, as shown in Fig. 5-15F. This capacitor is similar to Fig. 5-15E, except that the case is square instead of tubular. These capacitors are generally unavailable today, but can be replaced with other forms of electrolytic if you are not concerned with keeping the receiver in as near the original condition as possible.

For those who wish to keep the radio at least looking original, it is possible to carefully remove the outer case from the square capacitor, remove the innards, and then replace them with tubular electrolytics. When resealed, the capacitor will look very much like the original.

Solid-State Rectifiers

Although many antique radio buffs might consider the solid-state rectifier a modern development, it is actually quite old, at least in concept. Copper oxide rectifiers are a type of solid-state rectifier that dates back to the nineteenth century. These devices were copper disks that had one surface coated with copper oxide. Also common in older radios are rectifiers made of selenium.

Figure 5-16 shows the symbol for solid-state rectifiers and some common forms that these components might take. The symbol in Fig. 5-16A is universally recognized as representing semiconductor diodes, including rectifiers. The arrow points against the direction of electron flow. The arrowhead represents the anode (A), while the bar represents the cathode (K). An oddity of rectifier marking is that the cathode end, which is normally thought of as negative, is often marked with a "+" sign, as shown on those represented in Fig. 5-16.

The *selenium rectifier* is shown in Fig. 5-16B. This form of rectifier is found in many older radios, even though the rectifiers themselves are hard to locate today. The selenium rectifier consisted of a stack of metal plates that are each coated with selenium oxide.

The area of the plates indicates the current level, while the number of plates sets the voltage level. Typical selenium rectifiers were available in current ratings from 50 mA to several hundred milliamperes. Today, it is common practice to replace selenium rectifiers with silicon rectifier diodes, but more of that later.

The first silicon diode rectifiers available were the so-called *tophat* metal style shown in Fig. 5-16C. This rectifier was generally rated at 400 volts peak inverse voltage (PIV), and had a forward current rating of 500 mA. The large end of the top-hat rectifier is the cathode.

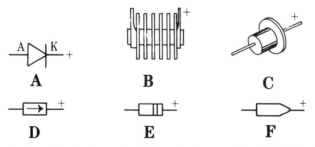

5-16   Rectifier diodes: (A) circuit symbol, (B) selenium rectifier, (C) 500 mA tophat, (D–F) epoxy-plastic.

Later, epoxy (plastic) rectifiers replaced the top-hat style. There are several popular styles of epoxy package, shown in Figs. 5-16D through 5-16F. The cathode end of these rectifiers is indicated by one or more of several means: *arrow, white band,* or *bullet-shaped end.*

The most popular series of silicon rectifiers is the 1N400X family. The "X" in the last digit will indicate the actual peak inverse voltage rating, starting with 1N4001 at 50 volts PIV. For radio service work, however, stick with the 1N4007 device. Like all 1N400X rectifiers, it is rated at 1 ampere forward current, but the PIV rating is 1000 volts.

Selecting Solid-State Rectifier Diodes

The two parameters that you will most often use to specify practical power-supply diodes are: the forward current and peak inverse voltage. Get these right, and in almost all cases the rectifier will work long and hard for you. The forward current rating of the diode must be at least equal to the maximum current load the power supply must deliver. That's common sense.

But in the real world a safety margin is also a necessity to account for tolerances in the diodes and variations of the real load, as opposed to the calculated load. It is also true that making the rating of the diode larger than the load current will greatly improve reliability. A good rule of thumb is to select a diode with a forward current rating of 1.5 to 2 times the calculated, or design goal, load current—or more if you can get it.

Although selecting a diode with a much larger forward current (e.g., 100 amperes for a 1-ampere circuit) is both wasteful and likely to make the diode not work like a diode, it is generally best to make the rating as high as feasible. The 1.5 to 2 times rule, however, should result in a reasonable margin of safety.

The peak inverse voltage rating can be a little more complicated. In unfiltered, purely resistive circuits, the PIV rating need only be greater than the maximum peak applied ac voltage (1.414 times rms). If a 20-percent safety margin is desired, then make it 1.7 times the rms voltage.

Most rectifiers are used in filtered circuits (e.g., Fig. 5-17A), and that makes the problem different. Figure 5-17B shows the simple capacitor filtered circuit redrawn to better illustrate the circuit action. Keep in mind that capacitor C is charged to the peak voltage (V_p) with the polarity shown.

The voltage across the transformer secondary (V) is in series with the capacitor voltage. When voltage V is positive, the transformer voltage and capacitor voltage cancel out, making the diode reverse voltage nearly zero. But when transformer voltage, V, is negative, the two negative voltages, V and Vc add up to twice the peak voltage, or approximately 2.83 times the rms voltage. Therefore, the absolute minimum PIV rating for the diode is 2.83 times the applied rms. If you prefer a 20-percent safety margin (a good idea), then make the diode PIV rating 3.4 times rms or more.

Older equipment that comes in for repair often uses hard-to-get or expensive vacuum tube rectifiers. We can use solid-state diodes to replace these tubes in almost all cases. The equivalent circuit is shown in Fig. 5-18A, while a practical

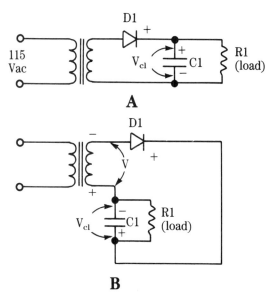

5-17 (A) Half-wave rectifier circuit with filter. (B) Circuit redrawn to show voltage relationships.

situation is shown in Fig. 5-18B. As shown here, for smaller rectifiers, we need only wire the solid-state diode across the pins of the rectifier tube socket.

Solid-state replacements for vacuum tube rectifiers can be made by mounting the solid-state diodes on an octal plug, or on a base salvaged from a burned out octal-based vacuum tube.

Selenium rectifiers are technically solid-state devices, but are no longer easily available. The standard technique in repairing power supplies based on the selenium rectifier is to use a silicon rectifier, such as the 1N4007 device. Some technicians like to mount the silicon diode on the defective selenium rectifier (Fig. 5-19A), but that is another bad practice.

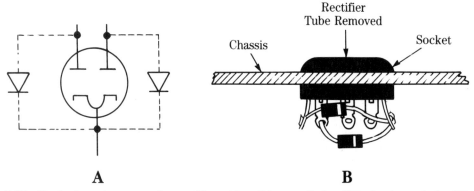

5-18 Replacing a vacuum tube rectifier with solid-state diodes: (A) circuit symbols, (B) actual installation.

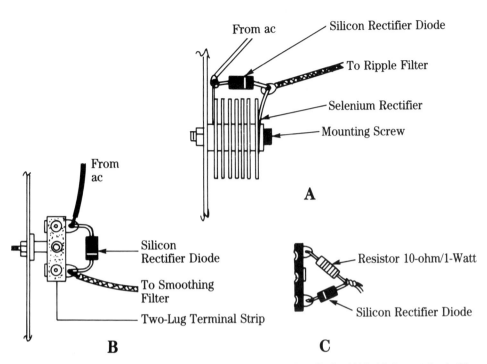

5-19   Replacing a selenium rectifier with a silicon rectifier diode: (A) bridging method, (B) terminal strip method, (C) with series resistor when three terminal strip not available.

The proper way is to remove the selenium rectifier, and then use its mounting hole in the chassis to install a two- or three-lug terminal strip (Fig. 5-19B). The silicon rectifier diode is connected between two *ungrounded* lugs on the terminal strip. Be careful to note whether or not a lug is grounded. It is common to find terminal strips on which the mounting tab is part of a terminal lug. Using a grounded lug will prove disastrous. Those who use a 10-ohm, 1-watt resistor in series with the silicon diode should either use a terminal strip with three insulated lugs, or use the connection scheme shown in Fig. 5-19C.

Voltage Regulators

The ac line voltage normally varies over a range of 105 volts to 125 volts, so we can also expect to see the rectified and filtered dc power-supply voltage to vary also. Unfortunately, some circuits are not tolerant of these variations. Some of the most sensitive of these are the local oscillators used in superheterodyne radios. These oscillators will drift in frequency with line-voltage variations, so stations will appear to wander across the radio's dial. The drift problem is especially troublesome on shortwave radios and FM broadcast receivers.

Another source of voltage variation in dc power supplies is changes in the load-current requirements. All dc power supplies have a certain internal resistance that drops a little of the open-circuit (no current) voltage when current is drawn. The actual output voltage is the open-terminal voltage less the loaded voltage. When the load current changes, then the voltage drop across the internal resistance also changes—and that changes the output voltage supplied to the radio receiver circuits.

The solution to the voltage-variation problem is to provide a *voltage regulator* in the circuit. Although modern solid-state radios use either zener diodes or three-terminal IC voltage regulators, older radios used gas-filled regulator tubes (Fig. 5-22A). You can identify the gas regulator tube by its blue, violet, or orange glow. These tubes are able to maintain a constant voltage drop between anode and cathode, despite variations in the applied voltage. They do so by varying their own internal resistance, which is a function of gas ionization.

These tubes have type numbers beginning with zero (not to be confused with older tubes, such as the 01A), and then a letter from the beginning of the alphabet. Typical type numbers are 0A2, OB2, and so forth. Gas regulator tubes are also sometimes labelled with the designation *VR-xx* in which the "xx" is the voltage-regulation value. For example, a VR-75 is a 75-volt regulator, a VR-105 is a 105-volt regulator, and a VR-150 is a 150-volt regulator. Some tubes are marked with both the type number and the VR-number. Thus, you might find an OA2 marked "0A2/VR-150."

The standard circuit for the gas regulator tube is shown in Fig. 5-22B. The voltage regulator is connected in shunt with the load (R_L) at the output of the dc power supply. A series resistor (R_s) is used to limit the output current. This resistor is a safety measure, needed because the regulator tube can become a very low resistance short circuit when the internal gas is ionized.

AC/DC Power Supplies

Transformers are expensive items compared to other radio receiver parts. As a result, manufacturers early on designed tranformerless "ac/dc" power supplies. These circuits will operate from either ac or dc power lines because they lack the ac-only transformer. Both vacuum tube and solid-state rectifiers are found in ac/dc power supplies, but in our example here we will use the solid-state form for simplicity.

Figure 5-20A shows the basic half-wave rectified dc power supply. The ac line is connected to the anode side of the rectifier (D1), and also to a *counterpoise ground*. This form of artificial ground is a wire bus that is used as the *de facto* ground for the radio. The negative side of the power supply, plus all RF, IF and audio signal return lines are connected to the artificial counterpoise ground.

Schematic diagrams for ac/dc radios might use two or three different ground symbols, depending on the situation. Although the symbols differ among certain old schematics, a common set is shown in Fig. 5-20B. The "garden rake" ground symbol represents a chassis ground. The counterpoise ground is represented by a small triangle, while the earth ground is represented by the three-line version of the symbol.

The switching for the ac/dc power supply is between the *neutral* line of the power mains and the counterpoise ground. This switching scheme is the most common encountered, but there is another type.

Figure 5-20C shows an extremely dangerous form of switch current. In this case, the chassis ground is used as the power-supply ground. While it serves the

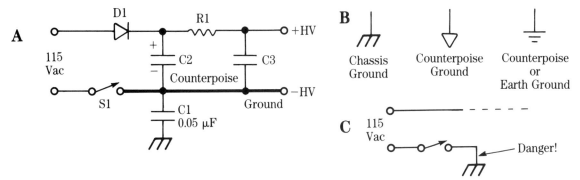

5-20 (A) ac/dc power supply using counterpoise ground. (B) Various circuit symbols used in diagrams. (C) Dangerous situation in some radios.

purpose when the receiver is buttoned up in a totally insulated cabinet, it becomes a real hazard that can kill people under at least to circumstances:

1. If the radio's plastic or wooden knobs are replaced with metal knobs
2. When the radio is out of the cabinet for repairs

The radio design of Fig. 5-20C depends on the ac power line being polarized to keep from making a hazardous situation. On some ac power lines one blade is larger than the other, and is intended to plug into the neutral line. But the safety of this arrangement depends on the power outlet on the wall being wired correctly all the way back to the pole, and there being no modification or replacement of the polarized plug.

The counterpoise ground is isolated from either the chassis ground or the earth ground (when the chassis is earth grounded), by a capacitor (C1) that has a low impedance at RF frequencies, but a high impedance at 60 Hz power line frequencies. This capacitor represents another hazard. Although there is no particular hazard when everything is working properly, a shorted bypass capacitor (C1) will make the radio similar to Fig. 5-20B.

Servicer's Safety Note: Repairing ac/dc radios can be very dangerous, especially when test equipment, work surfaces, or shop floors (including concrete floors) are grounded. These situations occur all too often, so consider ac/dc radios unsafe to service without specific measures being taken. If you service these radios, then it is very wise to use an *isolation transformer* to remove the ground reference factor from the ac power mains.

"All-American-Five" Radio Power Supplies

The "All American Five" is a standard mass-market radio design that uses five vacuum tubes: converter, IF amplifier, detector–AF preamplifier, audio power amplifier, and rectifier. Although the tube complement changed over the years, this design was very popular right up to the transistor era. Figure 5-21 shows a typical ac/dc power supply. In this earlier circuit the rectifier tube is the 35Z5, but keep in mind that others are also found (25Z5, 35W4, etc.).

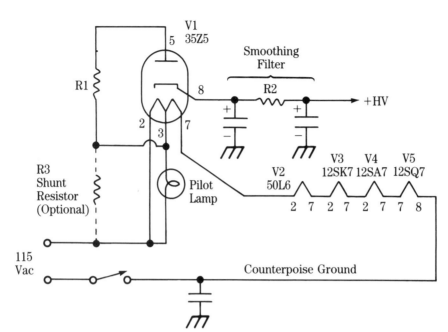

5-21 Rectifier circuit using a split-filament rectifier tube.

In the ac/dc All-American-Five radio, the filaments are connected in series, so their filament voltages must be added together. The circuit shown in Fig. 5-21 consists of the following filament voltages: 50, 35, 12, 12, and 12. These add to 121 volts. In some cases, the filaments add to a lower total voltage value in the 105- to 125-volt range, so a series "ballast" resistor is used to drop the remaining voltage.

Radios using the split-filament rectifier (25Z5, 35Z5, 35W4, etc.) use the radio pilot lamp to ballast one half of the rectifier filament, as shown in Fig. 5-21. This circuit sometimes leads to perplexing callbacks or immediate bench failures when the rectifier is replaced.

If the pilot lamp is bad, then the voltage drop across the rectifier filaments is excessive and the rectifier will fail prematurely. Whenever one of these rectifiers is replaced, be sure to check the pilot lamp too, before turning the set back on. Also, when the pilot lamp is found to be bad, be sure to check the health of the rectifier tube as well.

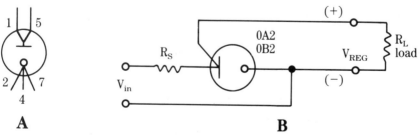

5-22 (A) gas voltage-regulator tube. (B) Regulator circuit.

6
The Audio Section

THE JOB OF THE AUDIO FREQUENCY (AF) SECTION OF THE RADIO RECEIVER IS TO BOOST THE small audio signal produced by the detector circuit to a level where it will competently drive an audio reproducer such as a loudspeaker or earphones. In this section we will take a look at audio amplifier circuits and audio reproducers typically used in radio receivers.

The audio section of the radio consists of at least an audio amplifier stage that drives a pair of earphones. These single-stage audio sections were used on early radios, even back in the earliest crystal set days. Later, especially when loudspeakers became available, the audio section became a little more complex (Fig. 6-1). Later audio stages, up to the present time, use at least two stages: first audio amplifier (also called *audio preamplifier*), and the second audio amplifier (also called *audio output* or *audio power amplifier*).

The last stage in the audio amplifier section is the reproducer, which is either a loudspeaker or a pair of earphones. Before studying the electronic circuits for the audio section let's first examine the two principal audio reproducers used in radios.

Audio Reproducers

Both earphones and loudspeakers are examples of *transducers*—devices that convert one form of energy to another. An *acoustical wave*, which is what the human ear responds to, is a pressure variation in air.

The reproducer, therefore, is a transducer that converts the ac electrical audio signals into air-pressure variations that the human ear can detect. Both earphones and loudspeakers could be called *electrically driven air pumps*.

Both types of reproducer are based on the electromagnetic principle, although some earphones are crystal devices, rather than electromagnets. Whenever an electrical current flows in a conductor, such as a wire, a magnetic field is generated around the wire.

If you form the wire into a coil (Fig. 6-2A), then the field is concentrated and increased by the mutual reinforcement of the windings with respect to each other. The strength of the magnetic field depends upon the level of current flowing in the coil: as current increases the magnetic field gets stronger.

The magnetic field is also made stronger if a suitable magnetic core is inserted into the center of the coil form. Although there will be a concentrated magnetic field surrounding the air-core coil, the field becomes much stronger if a core of iron, ferrite, or some other magnetic material is used for the coil. An *electromagnet* is a magnet formed by winding a coil over a suitable magnetic core, and this is the basis for sound reproducers.

The polarity (N/S) of the magnetic field in any given coil depends on the direction of current flow, which in turn is dependent upon the polarity of the battery. In Fig. 6-2 the coil is wound on a cylindrical form, and connected to a battery or other source of electrical current.

When current flows in the coil it is noted on a nearby compass that the compass needle deflects towards the coil. If the battery polarity is reversed, however, then the compass needle is deflected in the opposite direction. Clearly, the polarity of the magnetic field is a function of current direction.

The electromagnet, consisting of a coil of wire wound around a magnetic core, is capable of affecting more than simply a compass needle. In Fig. 6-2B we see a popular high school physics and electrical shop experiment used to illustrate this principle.

An electromagnet is placed beneath an iron disk, which is attracted by any form of magnet. The disk is suspended from a fixed point by a spring. When switch S1 is pressed, current flows in the coil windings, producing a magnetic field. The field attracts the magnetic iron disk, causing the spring to stretch an amount proportional to the strength of the electromagnet.

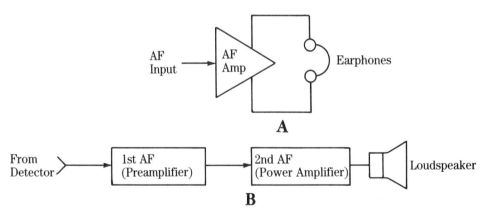

6-1 (A) Basic form of simple audio amplifier. (B) Complete audio section contains two stages—first AF and second AF.

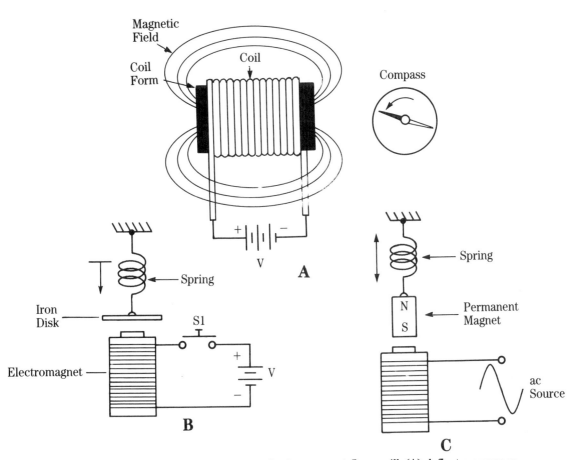

6-2 The magnetic field surrounding a coil when current flows will: (A) deflect a compass, (B) attract a mass, (C) attract or repel another magnet, depending on current polarity in the coil.

Now let's replace the iron disk of Fig. 6-2B with a small permanent bar magnet as shown in Fig. 6-2C. If the electrical current flowing in the electromagnet coil produces a North pole at the top end of the coil, then the suspended bar magnet is attracted to the electromagnet. But if the current is reversed so as to produce a South pole at the top end of the coil, then the bar magnet is repelled from the electro-magnet.

Guess what happens when, as in Fig. 6-2C, the battery is replaced with an alternating current source? The bar magnet will be alternately attracted to and then repelled from the electromagnet. It will oscillate up and down in step with the ac applied to the electromagnet coil.

Earphones

The earphone was perhaps the earliest form of sound reproducer used in radio receivers. The dynamic earphone was adapted from the telephone handset that was used at the time, and which is still essentially like modern handsets.

Earphones are still used today when one either wants to screen out room noise while listening to radio, not annoy others in the room who don't wish to listen to the radio, or when the output power available is insufficient to drive a loudspeaker. Modern portables use earphones for all three purposes.

Figure 6-3 shows the basic operation of the radio earphone. In the earphone, a coil for an electromagnet is wound around, and is concentric with, a permanent bar magnet. A magnetically sensitive metallic diaphragm, which is the air pumping surface, is placed at the end of the permanent magnet. When there is no current flowing in the coil, then the diaphragm is at rest (Fig. 6-3A).

When the current in the electromagnetic coil flows in one direction, the magnetic field of the coil reinforces the permanent magnet field, resulting in a stronger total field. The diaphragm is attracted to the magnet in this case (Fig. 6-3B).

When the current in the electromagnet coil is reversed, however, the magnetic field of the coil opposes the field of the permanent magnet, so the diaphragm is repelled to a distended position in the opposite direction (Fig. 6-3C). When an ac audio signal is applied to the electromagnet coil, the diaphragm will move back and forth as the electromagnet alternately adds to or opposes the magnetic field of the permanent magnet.

An example of earphone construction is shown in Fig. 6-3D. A U-shaped permanent magnet is placed in the protective covering, and the wires of the

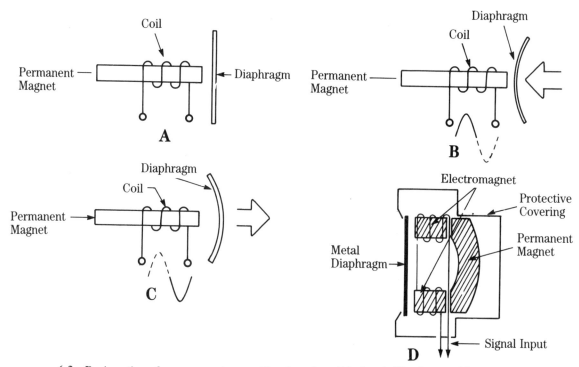

6-3   Basic action of a permanent magnet loudspeaker: (A) at rest, (B) when positive current flows, (C) when negative current flows, (D) basic structure of an earphone element.

electromagnet are wound around its arms. The metal diaphragm is placed across the ends of the magnet. For the usual "dual" earphones, one reproducer for each ear, there will be two earpieces connected in series.

Loudspeakers

A loudspeaker is a larger air pump. The metal diaphragm that was used to pump air in the earphones is replaced by a paper cone (Fig. 6-4) driven by a small electromagnet called a *voice coil* that interacts with either a permanent bar magnet or another electromagnet. There are two general categories of loudspeakers used in radios: *electromagnet* and *permanent magnet*.

The electromagnet speaker is shown in Fig. 6-5. An actual speaker from an old Stewart-Warner radio (c. 1938) is shown in Fig. 6-5A, while a schematic view of the speaker construction is shown in Fig. 6-5B. The main electromagnet consists of a soft iron-core wound with a large number of turns of wire that form the *field coil*. This structure forms the permanent magnet, even though it must be energized from an external source.

Coaxial to the field coil is another (weaker) electromagnet called the *voice coil*. The voice coil is wound around a bobbin at the center of the speaker cone (see the white circle at the center of the cone in Fig. 6-4). The voice coil will have a nominal impedance, such as 4, 8, or 16 ohms. A dc ohmmeter will show a certain resistance reading when the probes are placed across the voice coil.

The impedance can be approximated by a rule of thumb that the impedance of the speaker voice coil is 1.28 times the dc resistance. For that reason, you will

6-4   Loudspeaker cone serves as an "air pump."

A

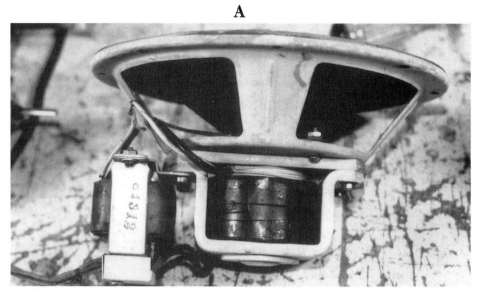

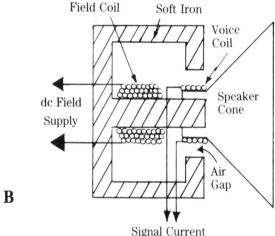

B

6-5   Electromagnet speaker: (A) form, (B) schematic.

sometimes see the same speaker listed as 3.2 ohms and at other times 4 ohms. In the former case it is the dc resistance that is listed, and in the latter case, the ac impedance at some test frequency.

Because the voice coil shares a core with the field coil, their magnetic fields will interact, forcing the voice coil, and hence the cone of the speaker, to move in and out in step with the ac audio signal applied to the voice coil.

Figure 6-6 shows the typical audio output section using an electromagnet speaker. The electromagnet field coil (FC) is energized by electrical current from the radio B+ power supply. It was the usual practice to use the FC to do double duty as the power-supply ripple-smoothing filter choke (see chapter 10 for an explanation of the power-supply ripple filter). The voice coil (VC) is connected to the audio output power-amplifier tube (V1), through an impedance-matching transformer (T1), which will be discussed more later.

The other coil in the circuit of Fig. 6-6 is the *hum-bucking* coil. It is used on many radios because the 60 Hz ac ripple hum present in the field coil would produce a hum in the loudspeaker. The hum-bucking coil is magnetically coupled to the field coil, but is connected in *series opposing* with the voice coil, so it will cancel the effects of field coil hum.

The *permanent-magnet (PM) speaker* is shown in Fig. 6-7. Conceptually, the PM speaker is identical to the electromagnet speaker, except that the field coil is replaced with a permanent magnet made of a cobalt–nickel alloy called *alnico*. The PM exerts a permanent field on the voice coil.

When ac audio signals flow in the voice coil, the magnetic field of the VC alternatively aids and then opposes the PM field, forcing the speaker cone to move in and out in step with the audio signal. No hum-bucking coil is needed on PM speakers.

The PM speaker has completely replaced the EM speaker in radio design. Although original replacement electromagnet speakers are still available from antique radio sources, some radios that originally used those speakers were converted many years ago to PM design. The voice coil is the same, so no audio circuit modifications are needed.

However, because the field coil is part of the dc power-supply ripple filter, either a series resistor or (better) a power-supply filter choke coil must be connected into the power-supply circuit to replace the missing field coil.

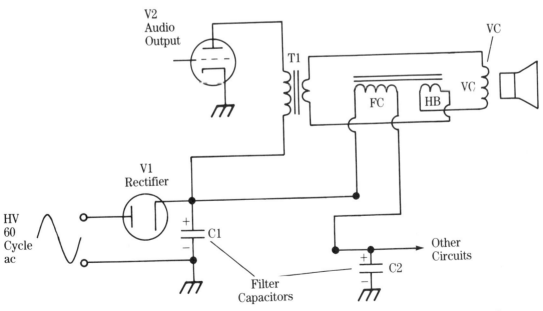

6-6 Circuit for an electromagnet loudspeaker when wired as part of the dc power-supply ripple filter (note the humbucking coil).

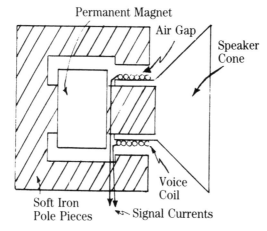

6-7 Schematic representation of the permanent magnet loudspeaker.

Audio Preamplifier Circuits

The job of the first audio stage is to serve as a preamplifier between the detector output and the AF power amplifier that actually drives the loudspeaker. The first audio stage is a linear voltage amplifier. Depending upon the radio, either triodes or pentodes are used for the first audio stage. Figures 6-8 and 6-9 show triode versions of these circuits; pentode versions are functionally the same, so they will not be covered separately.

Figure 6-8 shows the circuit for a transformer-coupled triode audio-preamplifier stage. Bias for the vacuum tube is supplied by a cathode-bias resistor (R1). The voltage drop across R1 created by the cathode current places the cathode at a slightly positive potential, so the grid is set slightly less positive (more negative) than the cathode. This meets the criteria for the "C" supply.

The capacitor shunted across the cathode bias-resistor is used to keep the cathode at ground potential for ac signal potentials, but above ground potential for dc. The capacitive reactance of C1 should be one-tenth of the resistance of R1 at the lowest frequency of operation for the stage.

For example, suppose the lower end -3 dB point for an audio amplifier is supposed to be 50 Hz, and R_1 is 1000 ohms. The reactance of C1 should then be 1000/10 or 100 ohms at 50 Hz. The value of C_1 will be:

$$C_{1\mu F} = \frac{1,000,000}{2 \pi F (R1/10)}$$

$$= \frac{1,000,000}{(2)(3.14)(50 \ Hz)(1000/10)}$$

$$= \frac{1,000,000}{31,400} = 32 \ \mu F$$

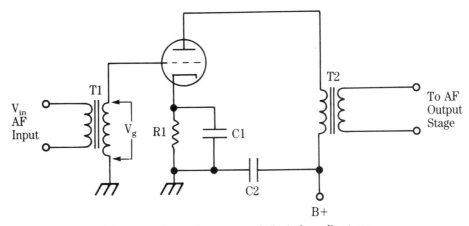

6-8   Typical transformer-coupled triode audio stage.

Capacitor C2 is used to decouple the B+ line, preventing audio signals from the power amplifier from affecting other stages. Capacitor C2 also sets the B+ end of the transformer at ground for ac signals while maintaining the B+ at a voltage high above ground.

Audio signals from earlier stages (such as the detector in radios) is coupled to the grid of V1 through transformer T1. The signal voltage at the grid (V_g) is proportional to input voltage, V_{in}, and the turns ratio of the transformer. The transformer turns ratio is the ratio of the number of turns in the secondary winding to the number of turns in the primary winding.

For example, suppose that the transformer has a 5:1 turns ratio. If a 5 mV signal is applied to the primary, then the grid voltage seen by the tube is 5 × 5 mV, or 25 mV. The power level of the signal is not increased, but the voltage level is stepped up five times.

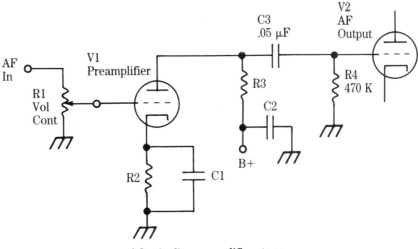

6-9   Audio preamplifier stage.

The output signal is coupled to the following stage through transformer T2. The primary impedance of this transformer is set to match the plate resistance of the first audio tube (V1), and the turns ratio is set to provide a higher voltage to the grid of the following stage if necessary.

The transformers used in Fig. 6-8 are expensive and take up a fair amount of space. Although they were used extensively in earlier radios, later models used resistance-capacitance (RC) coupling between stages. Figure 6-9 shows a first audio stage that is RC coupled to the AF output power amplifier

Three components are used in the coupling network: R3, C3 and R4. Resistor R3 is set to match the plate resistance of the first audio (preamplifier) tube, V1, while R4 is set to a much higher (10× or more) resistance than R3.

Capacitor C3 is used to prevent the high-voltage B+ supply applied to the plate of V1 from biasing the grid of the following stage. The capacitance of C3 is set to have a low reactance, relative to R3 and R4, at the lowest audio frequency that must be handled by the radio.

Resistor R1 in Fig. 6-9 is used as the volume control in this preamplifier circuit. The potentiometer is basically a voltage divider that selects the level of the signal voltage applied to the following preamplifier stage. In the next section we will look at several volume control schemes—some successful and some not—that have been used in radio receivers over the years.

Volume-Control Circuits

From almost the very earliest radio receivers, the volume control, used in a variety of circuits, has been one of two very similar components: a rheostat or a potentiometer. Both of these components are *variable resistors*.

Figure 6-10 shows the basic construction of both forms of volume control. In both cases a resistance element made of either resistance wire (Fig. 6-10A) or a carbon composition film (Fig. 6-10B), is mounted so that a wiper on a rotator shaft can be set to any point on the element, thereby selecting the desired resistance.

The principal difference between potentiometers and rheostats is that potentiometers have three terminals, one on each end plus the wiper, while the rheostat uses a wiper plus one end. The symbols for the two are nearly identical (Fig. 6-10C), but in the case of the rheostat, terminal number 3 is not used.

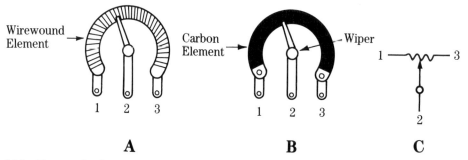

6-10 Forms of volume control: (A) wirewound element, (B) carbon element, (C) circuit symbol for potentiometer.

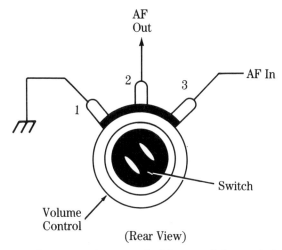

AF
Out

2
3

AF In

1

Switch

Volume
Control

(Rear View)

6-11 Wiring for potentiometer volume control to permit increasing sound level in the clockwise direction.

Figure 6-11 shows the normal convention for connecting volume-control potentiometers. When viewed from the rear, pin 1 is to the left. This pin is the ground or signal common terminal. When the left terminal (viewed from the rear) is used as the common, the volume control will cause the sound to get louder when the control is turned in a clockwise manner.

Also shown in Fig. 6-11 is a rear-mounted switch. Many volume controls include switches that are used for turning the radio on and off. This switch is connected to the 115 volt ac power line, so be careful when working around it.

One of the earliest methods of affecting the volume output of the radio receiver was to vary the filament voltage of the vacuum tube (Fig. 6-12). This method was used on the earliest radios, including Fleming diode detectors and DeForest triode circuits. A rheostat is placed in series with the "A" supply battery. Depending upon

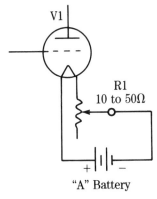

V1

R1
10 to 50Ω

+   −

"A" Battery

6-12 Crude volume control used on early radios adjusted directly heated cathode current from the "A" battery.

the filament current requirements, the full-scale value of the rheostat is normally something between 10 and 50 ohms.

This circuit did not work too well for several reasons. For one thing, the relationship between volume change and the rheostat setting is extremely nonlinear. Another problem was that turning down the filament voltage placed the tube in an inefficient mode of operation. There was also a threshold current, below which the tube will not work.

Figure 6-13 shows another volume-control circuit that was used in early radios. This circuit places a potentiometer in the primary of the antenna-coupler trans-former coil (T1). The antenna is connected to the wiper of the potentiometer, so the level of signal coupled to the tuned circuit is a function of the resistance setting of R1. Again, this method produced as many problems as it solved, so it was dropped early in the history of radio.

Another less than satisfactory method of volume control is shown in Fig. 6-14. In this circuit a rheostat is connected in series with the B+ dc power supply. The level of B+ voltage applied to the tube is set by the resistance of the rheostat. The defects included a nonlinear control over volume level, a bottom threshold, below which the rheostat could not go, and substantial distortion due to lowered B+, especially on strong signals.

Figure 6-15 shows a viable method that is still used for at least some purposes in radio receivers. The volume control in this case is a rheostat connected in series with the cathode resistor of the RF amplifier stage. By changing the bias on the tube, one can change the gain, and hence the output signal level.

This circuit was especially useful for certain very sensitive receivers or shortwave radio sets. In those radios the circuit of Fig. 6-15 was used most often for sensitivity, or RF gain, and will be used in conjunction with another circuit that serves as an audio volume control. The purpose of the sensitivity (RF gain) control is to reduce the gain of the RF amplifier on strong signals, thereby avoiding certain problems with overload from those strong local signals.

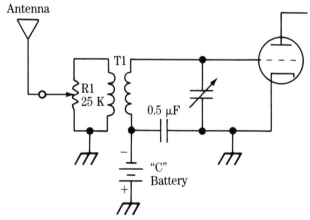

6-13   Antenna primary circuit volume control. It was crude, and only worked poorly. The same circuit recently showed up on a simple integrated circuit shortwave receiver kit.

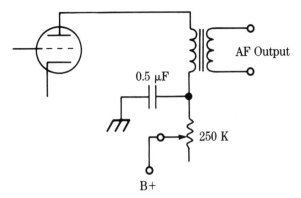

6-14 Volume control by adjusting the B+ potential.

Figures 6-16 and 6-17 show the two forms of the volume-control circuit that is most often used in radio receivers. In both cases the potentiometer is used as a voltage divider that selects a fraction of the available audio frequency signal, and then couples it to the grid of the first audio (preamplifier) stage.

In the case of Fig. 6-16, the potentiometer is shunted across the secondary of the detector output transformer. In Fig. 6-17, on the other hand, the potentiometer is connected as the load resistance of the diode detector circuit. The wiper on the potentiometer selects the fraction of the available detector output signal that will be sent to the first audio stage. A coupling capacitor (C1) is used to remove the dc offset potential that is normally produced by the diode detector.

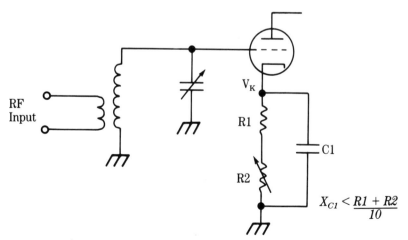

6-15 Cathode-bias volume control, sometimes called an RF gain control.

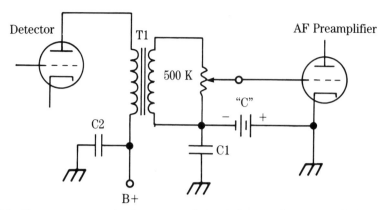

6-16 More conventional volume control circuit for transformer coupled radios.

Tone Controls

The *tone control* is used to custom-tailor the audio passband of the radio to whatever pleases the listener. Sometimes the frequency response is simply set according to personal preference. In other cases the tone is set according to the type of music being played. In still other cases, the high frequency treble range is "rolled off" to prevent noise signals from interfering with the audio signal being received.

The normal tone control used in most radios is the *treble roll-off* circuit. Fig. 6-18A shows a simple, nonadjustable method for rolling off the treble signal. In this case, a capacitor (C1) is placed between the plate of either the first audio, or audio output stage, and ground. Because capacitive reactance becomes lower as frequency increases, the capacitor will shunt more of the high-frequency signal to ground than low-frequency bass signals.

A variable treble roll-off tone-control circuit is shown in Fig. 6-18B. In this circuit a rheostat is connected in series with the treble roll-off capacitor, producing either more or less roll-off depending upon its setting. This resistor is usually a front

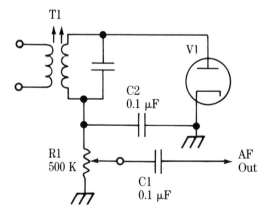

6-17 Volume control circuit for RC-coupled radios.

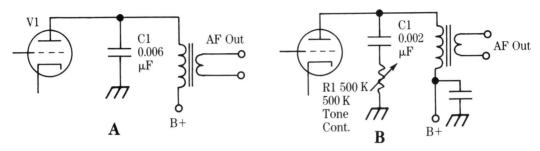

6-18 Tone controls: (A) simple tone tailoring uses a plate bypass capacitor, (B) RC network forms variable tone control.

panel control, and will be marked TONE or TONE CONTROL.

A *base-treble* control circuit is shown in Fig. 6-19. In this circuit there are two frequency-response tailoring networks: C1 and C2/L1. The potentiometer selects how much of each of these networks enters the circuit. When the potentiometer is balanced in the center of its range, the response is said to be *flat*. The bass and treble functions are determined by the setting of R1.

Audio-output stages

The audio-output stage is an AF power amplifier that develops the power level needed to drive the loudspeaker. The AF power amplifier receives the amplified signal from the first audio, and then amplifies it further. The difference between the first-audio preamplifier and the output amplifier is that the preamplifier is a voltage-amplifier, while the output amplifier is a power-amplifier.

Classes of audio power amplifier

The amplifier classification system used in radio circuits divides amplifiers into four classes: *A*, *B*, *C*, and a combination class *AB*. The class-C amplifier is not suitable to audio amplification, and is only used in certain RF circuits in transmitters. The

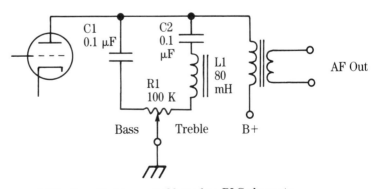

6-19 Bass-Treble control based on RLC elements.

distortion factor is so high in class-C amplifiers that it cannot be used in either audio circuits or those RF circuits where linear amplitude is important.

The basis for the classification system is *plate conduction angle*. The conduction angle is the duration of plate current flow relative to the input sinewave cycle. In class-A amplifiers, plate current flows throughout the entire 360 degrees of the input sinewave.

These stages tend to be biased approximately midway on the plate characteristic curve, so that input signal voltages do not force the plate either into saturation or cutoff on positive or negative peaks respectively. The class-A amplifier tends to be inefficient, so only 40 to 50 percent of the dc power consumed from the power supply is delivered to the speaker as available audio power. The remainder of the power is given off as heat.

In class-B audio amplifiers, the plate current only flows over 180 degrees of the input signal. Because of this fact, the output will be distorted unless a pair of tubes is used, working together in a *push-pull* circuit, to produce the entire 360-degree output signal. In that type of amplifier, one tube is used for each half of the sinewave cycle. The theoretical maximum efficiency of the class-B amplifier is on the order of 73 percent, but in reality 50 to 60 percent is more likely.

The class AB is a related class, used in some radios. In these amplifiers, the plate current flows for more than 180 degrees, but less than 360 degrees. Class-AB amplifiers tend to be somewhat more efficient than either class-A or class-B circuits.

Class-A power amplifiers

The class-A power amplifier consists of a single tube that conducts plate current over the entire 360 degrees of a sinewave input signal. Because only one tube is used in this circuit, the class-A power amplifier is said to be *single-ended*. In Fig. 6-20A, tube V1 is a class-A preamplifier (voltage amplifier), while V2 is a class-A audio power amplifier.

Bias for the tube is provided by a cathode resistor (R6) that is bypassed with a capacitor that sets the cathode at ground potential for ac signals. The value of R6 depends upon the tube type and the power level, but in general it will be between 100 ohms and 600 ohms in common radio power-amplifier circuits. The capacitance of C5 is set so that its reactance at the lowest audio frequency will be R6/10.

Input signals can be coupled to the control grid of V2 either by transformer, or by the RC coupling shown in Fig. 6-20A. This signal is received from the plate of the first audio preamplifier stage, V1.

In any electrical system the greatest power transfer occurs between a source (such as the power-amplifier tube) and a load (such as a loudspeaker) when the impedances of the two are matched. Unfortunately, the plate resistance (R_p in Fig. 6-20B) of a vacuum tube is very much larger than the impedance of the loudspeaker.

For example, a typical value for the plate resistance is 2,500 ohms, while for the speaker it is 4 ohms. This tremendous mismatch would cause all manner of grief and mischief in the radio without the *audio output transformer* (T1), used to match the plate resistance to the speaker impedance.

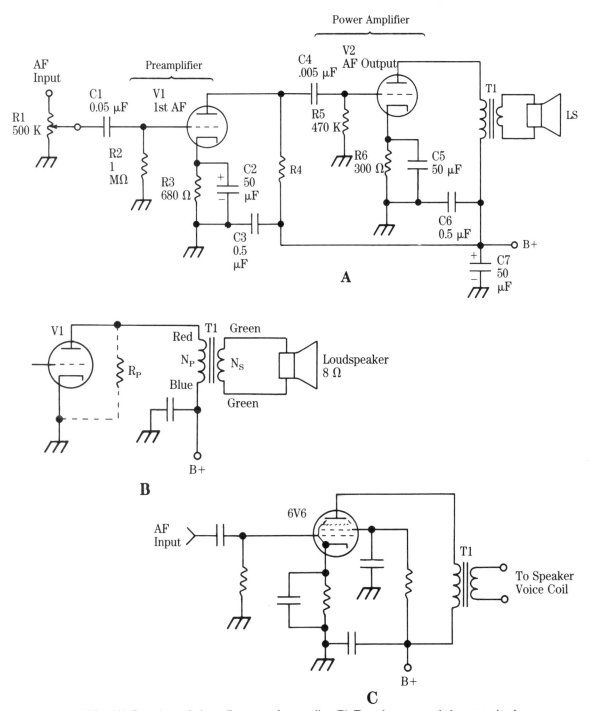

6-20 (A) Complete triode audio stages for a radio. (B) Transformer coupled output circuit matches the low impedance loudspeaker to the high impedance plate resistance. (C) Beam power tube AF output stage.

Audio output transformers are specified in either or both of two ways. First, they may list the two impedances. For example, one transformer might list the impedance of the primary as 2,500 ohms, and the voice coil (often marked "V.C.") impedance as 4 ohms. The same transformer could also be listed according to the *turns ratio* (N), the ratio of the number of turns in the primary winding (N_p) to the number of turns in the secondary winding (N_s), or $N = N_p/N_s$. The impedance transformation ratio is:

$$\frac{N_p}{N_s} = \sqrt{\frac{Z_p}{Z_s}}$$

Where: N_p = number of turns in the primary winding
N_s = number of turns in the secondary winding
Z_p = "reflected impedance" looking into the primary winding
Z_s = impedance connected to the secondary winding

The *reflected impedance* is the load seen by the power-amplifier tube, and should match as nearly as possible the plate resistance of the tube.

The triode circuit of Fig. 6-20A was replaced prior to World War II by a power-pentode circuit such as Fig. 6-20C. Tubes such as 6V6, 12V6, 6AQ5, and other beam power tubes were used for the power-amplifier stage. They gave a superior linearity and better power sensitivity.

One of the aspects of the class-A power amplifier is that power flows over the entire 360 degrees of a sinewave input cycle. As a result, one advantage of the class-A amplifier is that a single tube can be used for the power amplifier.

The down side, however, is that there is a lot of wasted heat in a class-A power amplifier. The total dc power consumed by a class-A amplifier is constant, regardless of the applied signal level. This power is divided up between audio power delivered to the speaker, and anode heat dissipated by the tube.

As a consequence, the plate current drawn by the tube will be constant regardless of signal level. Thus, the tube operated as a class-A amplifier will run hot because of the excess heat given up by the anode. In contrast to the inefficiency of the single-ended class-A amplifier, the push-pull class-B or class-AB amplifier is considerably more efficient.

Push-pull class B and class AB audio power amplifiers

Figure 6-21 shows the push-pull audio power-amplifier circuit. It uses two triode power-amplifier tubes arranged in a balanced circuit. Although triodes are used in our example, beam-power pentodes were also used extensively in this type of circuit. The audio output transformer (T2) is center-tapped, with the B+ power supply connection made to the center-tap. The two ends of the audio output transformer are connected to the plates of the two power-amplifier tubes.

If an equal current flows in the two tubes, this connection causes equal currents (I_1 and I_2) to flow in the primary winding of T2. But notice that these currents flow in opposite directions in the primary, so they will cancel out each other.

The input transformer uses a single-ended primary winding, but a center-tapped

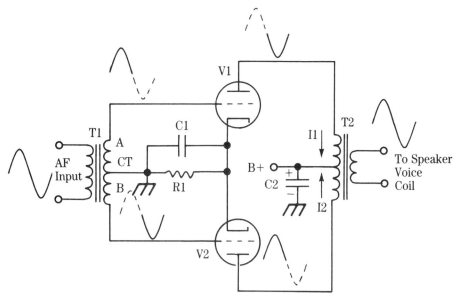

6-21 Transformer coupled triode push-pull AF power amplifier circuit.

secondary winding. Because the center-tap is grounded, a sinewave applied to the primary causes equal but opposite polarity voltages to appear at the ends (A and B). These potentials are used to drive the grids of power-amplifier tubes V1 and V2. Thus, the two tubes are driven 180 degrees out of phase with each other.

When the signal applied to the grid of V1 is positive, as shown, the signal at the grid of V2 has the same voltage but is of the opposite polarity (it is negative).

The tubes in a push-pull amplifier are biased by a cathode resistor to the cutoff point. When a negative signal is applied to the grid, the tube is driven even further into the cutoff region. When the signal is positive, the tube is driven out of cutoff into conduction.

In Fig. 6-21 the grid of V2 is negative, so it is cut off. The grid of V1 sees a positive signal, so it will conduct current and cause the amplified half-cycle to appear in the plate circuit.

This half-cycle signal flows in the primary of the output transformer, and is therefore coupled to the load. On the alternate half-cycle, V1 is now cut off and V2 is conducting. The alternative half-cycle is amplified and coupled to the load through the output transformer. The signal across the load is the sum of both half-cycles, so it represents the entire input signal reconstructed.

The push-pull amplifier offers both greater output power and better efficiency than the single-ended class-A amplifier. In addition, the action of the push-pull amplifier tends to cancel even-order harmonics, so the push-pull amplifier has better linearity, or *fidelity*, than single-ended amplifiers.

The interstage transformer between the first audio stage and the push-pull power amplifier is expensive, takes up a lot of space, and offers a certain amount of distortion in its own right if the impedance isn't flat across the audio spectrum. Some radios use a *phase inverter* circuit (Fig. 6-22). This circuit works because the signal across the cathode resistor of the triode (V1) is in phase with the input signal, while the plate signal is 180 degrees out of phase with the input signal. Thus, on any given half-cycle the signals applied to the grids of the push-pull power amplifier tubes (V2 and V3).

Screen grid feedback circuits

A small amount of negative (out of phase) feedback is used in many audio power-amplifiers to improve linearity and reduce noise. Most feedback signals are derived from the secondary of the output transformer and are applied to the first audio preamplifier stage.

However, there are at least two methods that apply negative feedback from the output transformer to the screen grid. In Fig. 6-23A, the primary of the audio output transformer is tapped (but not center-tapped) with the B+ power-supply connection being made to the tap. The small section of winding between the center tap and the screen-grid end of the transformer places a small audio signal, of opposite polarity to the plate signal, in series with the dc voltage that is applied to the screen grid.

Another circuit is shown in Fig. 6-23B. This push-pull amplifier, known as the *ultralinear push-pull audio power amplifier*, uses a special audio output transformer that has a pair of screen-grid taps on the primary winding. The screen grids therefore receive both the dc potential that is applied to the plates and a small sample of the ac signal that appears across the transformer primary. This ac signal on the screen grid represents a negative feedback signal.

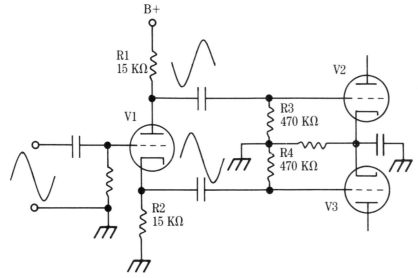

6-22   Push-pull power amplifier that uses a phase inverter stage instead of an audio input transformer.

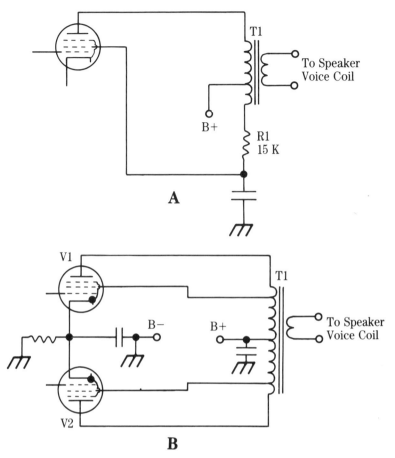

6-23 (A) Single-ended stage that uses negative feedback to improve distortion. (B) Ultralinear push-pull amplifier configuration.

Replacing Volume, Tone, or RF Gain Controls

The volume control, gain control, and (if used) RF gain controls on a radio are either potentiometers, or on the very oldest sets, rheostats. New controls can be bought at radio-TV parts wholesalers. You will need to know some facts before buying, however:

- What is the resistance? Measure across the two outer terminals.
- Is it wirewound or carbon? Look at the element—most are carbon.
- Is the resistance vs. shaft-angle taper audio or linear? Use audio for volume controls and linear for the others.
- Does the volume control have a loudness tap? Look on the underside for a "spare" terminal.

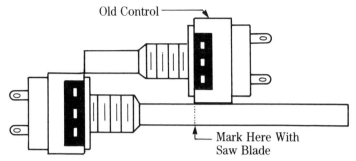

6-24   Measuring a "universal" potentiometer shaft for replacement in a radio.

- Is the shaft half-round or full-round? Is the shaft metal or made of an insulated material? Is it smooth or splined?
- Is there a switch mounted on the rear of the control? If so, how many terminals.

Once you have selected a control, you will probably find that it has a shaft that is too long (a consequence of universality). Figure 6-24 shows how to measure the shaft of the new control against the old control. Mark the shaft of the new control at the point shown with a hacksaw blade or scribe. Next, place the control in a bench vise and cut off the shaft at the marked point.

Replacing Multiresistors

Many older radios had multiple-resistor packages in which two or more series resistors are inside one package. These resistors are basically one long resistance element tapped at the desired points. They are often shown in schematics as separate resistors, although some are shown as an assembly.

When one section goes open, it can be bridged with a separate power resistor, as shown in Fig. 6-25. Two methods of mounting are shown in Fig. 6-25, so select the one that seems reasonable at the time.

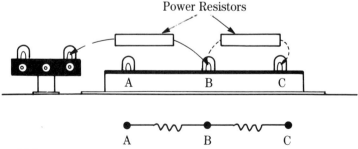

6-25   One method for replacing one section of multitap resistor.

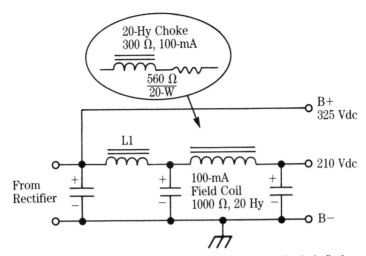

6-26 Method for replacing the loudspeaker field coil "power supply choke" when converting from EM to PM speaker.

Replacing Electromagnet Loudspeakers With Permanent Magnet Loudspeakers

The first loudspeakers used in radios had an electromagnet instead of a permanent magnet. This electromagnet, called the *field coil*, was often used as a power-supply ripple-filter smoothing choke for the dc power supply (Fig. 6-26). When one of these speakers is replaced, the permanent magnet speaker can be used directly with the audio output transformer, but removing the choke open-circuits the dc power-supply.

This field coil can be replaced with a chassis-mounted choke coil similar to L1 in Fig. 6-26. However, if the dc resistance of the field coil is higher than the choke dc resistance (likely), then also add a power resistor in series with the choke that makes up the difference (at least approximately).

7

The Radio Front-End: RF Amplifiers, First Detectors, and IF Amplifiers

THE RADIO FRONT-END INCLUDES ALL THE CIRCUITRY FROM THE ANTENNA TO THE detector, including the IF amplifiers. Some authors, incidentally, do not include the IF amplifiers in the definition of *front-end*. But because the IF amplifier and the RF amplifier share certain characteristics (both are tuned radio frequency amplifiers), I deemed it appropriate to include the entire radio prior to the second detector in the operational definition of *front-end* for purposes of this book.

In this chapter we will look at a number of topics and circuits that pertain to the radio front-end. Because the heart of the front-end is the tuned resonant circuit, we will cover these circuits in detail early in the chapter. Following that discussion we will look at RF amplifiers, mixers, local oscillators, converters, and IF amplifier circuits.

Tuned resonant circuits

Tuned resonant circuits, also called *tank circuits* or *LC circuits*, are used in the radio front-end to select from the myriad of stations available at the antenna. The tuned resonant circuit is made up of two principal components: inductors, and capacitors (also known in old radio manuals as *condensers*).

In this section we will examine inductors and capacitors separately, and then in combination to determine how they function together to tune the radio's RF, IF, and LO circuits. But first, we need to make a brief digression to discuss *vectors* because they are used in describing the behavior of these components and circuits.

Vectors

A vector (Fig. 7-1A) is a graph device that is used to define the *magnitude* and *direction* (both are needed) of a quantity or physical phenomenon. The length of the

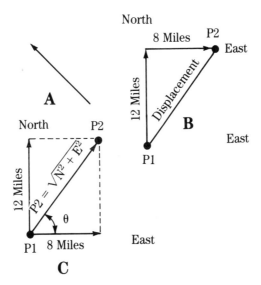

7-1 Vector notation: (A) A vector has *length* and *direction*. (B) A journey that can be represented by vectors. (C) Vector representation of journey shows that the resultant (P1 ≥ P2) can be calculated from the individual vectors.

arrow defines the magnitude of the quantity, while the direction in which it is pointing defines the direction of action of the quantity being represented.

Vectors can be used in combination with each other. For example, in Fig. 7-1B we see a pair of displacement vectors that define a starting position, P1, and a final position, P2, for a person who traveled from point P1, 12 miles north and then 8 miles east to arrive at point P2.

The *displacement* in this system is the hypoteneus of the right triangle formed by the "north" vector and the "east" vector. This concept was once illustrated pungently by a university bumper sticker's directions to get to a rival school: "North 'til you smell it, east 'til you step in it."

Figure 7-1C shows a calculations trick with vectors that is much used in engineering, science, and especially electronics. We can *translate* a vector parallel to its original direction, and still treat it as valid. The "east" vector (E) has been translated parallel to its original position so that its tail is at the same point as the tail of the "north" vector (N). This allows us to use the Pythagorean theorem to define the vector. The magnitude of the displacement vector to P_2 is given by:

$$P_2 = \sqrt{N^2 + E^2}$$

But recall that the magnitude only describes part of the vector's attributes. The other part is the direction of the vector. In the case of Fig. 7-1C, the direction can be defined as the angle between the "east" vector and the displacement vector. This angle (<) is given by:

$$< \, = cos^{-1}\,(E/P_2)$$

In generic vector notation there is no "natural" or "standard" frame of reference, so the vector can be drawn in any direction, so long as the users understand what it means. In the preceeding system, we have adopted, by convention, a method that is basically the same as the old fashioned Cartesian coordinate system X-Y graph. In the example of Figs. 7-1B and 7-1C the "X" is the "east" vector, while the "Y" is the "north" vector.

In electronics, vectors used to describe voltages and currents in ac circuits are standardized (Fig. 7-2). They use this same kind of cartesian system in which the inductive reactance (X_L), i.e., the opposition to ac exhibited by inductors, is graphed in the "north" direction, the capacitive reactance (X_c) is graphed in the "south" direction, and resistance (R) is graphed in the "east" direction.

Inductance and inductors

Inductance (L) is a property of electrical circuits that opposes change in the flow of current. Note that word "changes;" it is important. As such, it is somewhat analogous to the concept of inertia in mechanics. An inductor stores energy in a magnetic field (a fact which we will see is quite important). In order to understand the concept of inductance we must understand three physical facts:

1. When an electrical conductor moves relative to a magnetic field, a current is generated, or "induced," in the conductor. An *electromotive force* (EMF, or "voltage") appears across the ends of the conductor.

2. When a conductor is in a magnetic field that is changing, a current is induced in the conductor. As in the first case, an EMF is generated across the conductor.

3. When an electrical current moves in a conductor, a magnetic field is set up around the conductor.

According to *Lenz's law*, the EMF induced into a circuit is ". . . in a direction that opposes the effect that produced it." From this fact we can see the following effects:

1. A current induced by either the relative motion of a conductor and a magnetic field, or changes in the magnetic fields, always flows in the

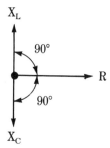

7-2   Vectors used to represent RLC circuits. Convention places the resistance vector (*R*) toward the east, the inductive reactance vector (X_L) toward the north, and the capacitance reactance vector (X_C) towards the south.

direction that sets up a magnetic field that opposes the original magnetic field.

2. When a current flowing in a conductor changes, the magnetic field that it generates changes in a direction that induces a further current into the conductor that opposes the current change that caused the magnetic field to change.

3. The EMF generated by a change in current will have a polarity that is opposite the polarity of the potential that created the original current.

The unit of inductance (L) is the *henry* (H). The accepted definition of the henry is the inductance that creates an EMF of one volt when the current in the inductor is changing at a rate of one ampere per second, or mathematically:

$$V = L \frac{\Delta I}{\Delta t}$$

Where: V = induced EMF, in volts
L = inductance, in henrys
I = current, in amperes
t = time, in seconds
Δ = a small change

The henry is the appropriate unit for inductors such as the smoothing-filter chokes (chapter 5) used in dc power supplies, but is too large for RF and IF circuits. In those circuits the sub-units, *millihenrys* (mH) and *microhenrys* (μH) are used. These are related to the henry by: 1 henry = 1,000 millihenrys = 1,000,000 microhenrys. Thus, 1 mH = 10^{-3} H and 1 μH = 10^{-6} H.

One of the phenomena just listed is called *self-inductance*. When the current in a circuit changes, the magnetic field generated by that current change also changes. This changing magnetic field induces a counter-current in the direction that opposes the original current change. This induced current also produces an EMF (discussed above), called the *counter electromotive force* (CEMF). As with other forms of inductance, self-inductance is measured in henrys and its subunits.

Self-inductance can be increased by forming the conductor into a multiturn coil (Fig. 7-3A) in such a manner that the magnetic fields in adjacent turns reinforce each other. This means that the turns of the coil must be insulated from each other. The coil wound in this manner is called an inductor, or simply a coil, in RF/IF circuits.

The inductors pictured in Figs. 7-3A and 7-3B are called solenoid wound coils if the length (l) is greater than the diameter (d). The inductance of the coil is actually self-inductance, but the "self-" is usually dropped in favor of simply "inductance."

Several factors affect the inductance of a coil. Perhaps the most obvious are the length, the diameter, and the number of turns in the coil. Also affecting the inductance is the nature of the core material and its cross-sectional area. In the examples of Fig. 7-3, the cores are simply air, and the cross-sectional area is directly related to the diameter, but in many radio circuits the core is made of powdered iron or ferrite materials.

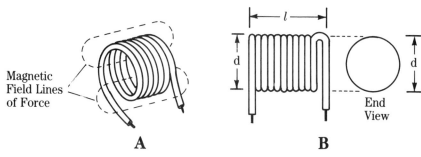

Magnetic
Field Lines
of Force

End
View

A   **B**

7-3   All current-carrying wires have a magnetic field: (A) forming the wire into a coil enhances the field, (B) a solenoid-wound coil.

For an air-core solenoid wound-coil in which the length is greater than 0.4d, the inductance can be approximated by:

$$L_{\mu H} = \frac{d^2 N^2}{18d + 401}$$

The core material has a certain *magnetic permeability* (μ), which is the ratio of the number of lines of flux produced by the coil with the core inserted, to the number of lines of flux with an air core (i.e., core removed). The inductance of the coil is multiplied by the permeability of the core.

Combining inductors

When inductors are connected together in a circuit, their inductance combines similar to the resistance of several resistors in parallel or series. For inductors in which their respective magnetic fields do not interact:

A) Series-connected inductors:

$$L_{total} = L1 + L2 + L3 + \ldots + L_n$$

B) Parallel-connected inductors:

$$L_{total} = \frac{1}{\dfrac{1}{L1} + \dfrac{1}{L2} + \dfrac{1}{L3} + \ldots + \dfrac{1}{L_n}}$$

Or, in the special case of two inductors in parallel:

$$L_{total} = \frac{L1 \times L2}{L1 + L2}$$

If the magnetic fields of the inductors in the circuit interact, then the total inductance becomes somewhat more complicated to express. For the simple case of two inductors in series, the expression would be:

A) Series inductors:

$$L_{total} = L1 + L2 \pm 2M$$

M is the *mutual inductance* caused by the interaction of the two magnetic fields. (Note: $+M$ is used when the fields aid each other, and $-M$ is used when the fields are opposing.)

$$L_{total} = \cfrac{1}{\cfrac{1}{L1 \pm M} + \cfrac{1}{L2 \pm M}}$$

Radios made in the 1920s through the early 1930s used air-core coils in their tuning circuits (Fig. 7-4). Note that two of the coils in Fig. 7-4 are aligned at right angles to the other one. The reason for this arrangement is not mere convenience, but is rather a tactic used by the radio designer to prevent interaction of the magnetic fields of the respective coils. In general, for coils in close proximity to each other:

1. Maximum interaction between the coils occurs when the coils' axes are parallel to each other;
2. Minimum interaction between the coils occurs when the coils' axes are at right angles to each other.

For the case where the coil axes are along the same line, the interaction depends on the distance between the coils.

7-4 Inductors in this TRF radio are placed at right angles to each other to prevent cross-coupling between them.

Adjustable coils

There are several practical problems with the standard fixed coil just discussed. For one thing, the inductance cannot easily be adjusted either to tune the radio or to trim the tuning circuits to account for the tolerances in the circuit.

Air-core coils are difficult to adjust. They can be lengthened or shortened, the number of turns can be changed, or a tap or series of taps can be established on the coil in order to allow an external switch to select the number of turns that are allowed to be effective. None of these methods is terribly elegant, though all have been used in one application or another.

The solution to the adjustable inductor problem, developed relatively early in the history of mass-produced radios, and still used today, is to insert a powdered iron or ferrite core (or "slug") inside the coil form (Fig. 7-5). The permeability of the core will increase or decrease the inductance according to how much of the core is inside the coil. If the core is made with either a hexagonal hole or screwdriver slot, then the inductance of the coil can be adjusted by moving the core in or out of the coil. These coils are called *slug-tuned inductors*.

Inductors in ac circuits

Impedance (Z) is the *total opposition to the flow of alternating current (ac) in a circuit*. It is analogous to resistance in dc circuits. The impedance is made up of a resistance component (*R*) and a component called *reactance* (*X*). Like resistance,

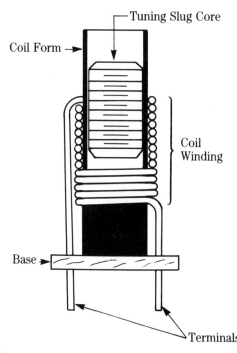

7-5   "Slug-tuned coils" offer variable inductance by inserting a threaded powdered-iron core (or "slug") into the coil.

reactance is measured in ohms. If reactance is produced by an inductor, then it is called inductive reactance (X_L), and if it is produced by a capacitor it is called capacitive reactance (X_c). Inductive reactance is a function of the inductance and the frequency of the ac source:

$$X_L = 2\,\Pi\,f\,L$$

Where: X_L = inductive reactance in ohms
$\quad\quad f$ = ac frequency in hertz (Hz)
$\quad\quad L$ = inductance in henrys (H)

In a purely resistive ac circuit (Fig. 7-6A), the current (I) and voltage (V) are in phase with each other; i.e., they rise and fall at exactly the same times in the ac cycle. In vector notation (Fig. 7-6B), the current and voltage vectors are along the same axis, which is an indication of the zero-degree phase difference between the two.

In an ac circuit that contains only an inductor (Fig. 7-7A), and is excited by a sine-wave ac source, then a change in current is opposed by the inducatance. As a result, the current (I) in an inductive circuit lags behind the voltage (V) by 90 degrees. This is shown with vectors in Fig. 7-7B, and as a pair of sinewaves in Fig. 7-7C.

The ac circuit that contains a resistance and an inductance (Fig. 7-8A) shows a phase shift (Δ), shown vectorilly in Fig. 7-8B, that is other than the 90 degrees seen in purely inductive circuits. The phase shift is proportional to the voltage across the inductor and the current flowing through it. The impedance of this circuit is found by the Pythagorean rule described earlier, also called the *root of the sum of the squares* method (see Fig. 7-8):

$$Z = \sqrt{R^2 + (X_L)^2}$$

The coils used in radio receivers come in a variety of different forms and types, but all radios except the very crudest untuned crystal sets will have at least one coil. Now let's turn our attention to the other member of the LC tuned circuit.

Capacitors and capacitance

Capacitors, also called *condensers* in early tests, are the other components used in radio tuning circuits. Like the inductor, the capacitor is an energy storage device.

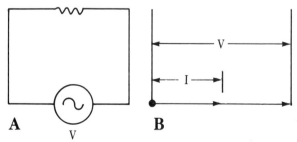

7-6 (A) Resistor circuit. (B) Vector representation of current (I) and voltage (V).

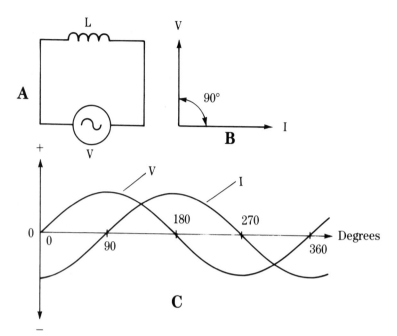

7-7   (A) Inductor circuit. (B) Vector relationships. (C) Phase relationships.

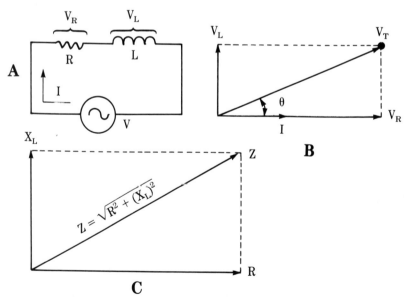

7-8   (A) Resistor-inductor (RL) circuit. (B) Vector relationships (resultant is impedance). (C) Solving for impedance (Z).

While the inductor stores electrical energy in a magnetic field, the capacitor stores energy in an *electrical* (or *electrostatic*) field. Electrical charge (Q) is stored in the capacitor.

The basic capacitor consists of a pair of metallic plates facing each other, and separated by an insulating material called a *dielectric*. This arrangement is shown schematically in Fig. 7-9A, and in a real configuration in Fig. 7-9B. The fixed capacitor shown in Fig. 7-9B consists of a pair of square metal plates separated by a dielectric.

Although this type of capacitor is not terribly practical, it was once used quite a bit in transmitters. Spark transmitters of the 1920s often used a glass and tinfoil capacitor fashioned very much like Fig. 7-9B. Layers of glass and foil were sandwiched together to form a high-voltage capacitor. A 1-foot-square capacitor made of 1/8-inch-thick glass and foil has a capacitance of about 2,000 pF.

Units of capacitance

The *capacitance* (C) of the capacitor is a measure of its ability to store current, or more properly, electrical charge. The principal unit of capacitance is the *farad*, named after physicist Michael Faraday. One farad is the capacitance that will store one coulomb of electrical charge (6.28 × 10^{18} electrons) at an electrical potential of one volt. Or, in math form:

$$C_{\text{farads}} = \frac{Q_{\text{coulombs}}}{V_{\text{volts}}}$$

The farad is far too large for practical electronics work, so subunits are used. The *microfarad* (μF or mF) is 0.000001 farad (1 F = 10^6 μF). The *picofarad* (pF) is 0.000001 μF, or 10^{-12} farads. In older radio texts and schematics, the picofarad was called the *micromicrofarad* (μμF or mmF).

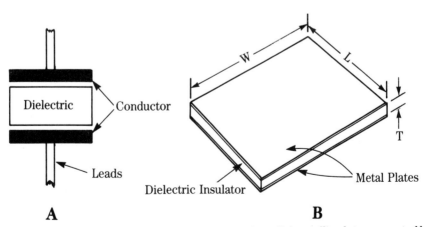

A   **B**

7-9 A capacitor (or "condenser") consists of a pair of parallel metallic plates separated by an insulator called a *dielectric*. (A) Symbolic form, (B) physical form for a school experiment capacitor.

The capacitance of the capacitor is directly proportional to the area of the plates (in terms of Fig. 1B, L×W), inversely proportional to the thickness (T) of the dielectric (or the spacing between the plates, if you prefer), and directly proportional to the *dielectric constant* (K) of the dielectric.

The dielectric constant is a property of the insulator material used for the dielectric. The dielectric constant is a measure of the material's ability to support electric flux, and is thus analogous to the permeability of a magnetic material. The standard of reference for dielectric constant is a perfect vacuum, which is said to have a value of $K = 1.00000$. Other materials are compared with the vacuum. The values of K for some common materials are:

Vacuum	1.0000
Dry air	1.0006
Paraffin (wax) paper	3.5
Glass	5 to 10
Mica	3 to 6
Rubber	2.5 to 35
Dry wood	2.5 to 8
Pure (distilled) water	81

The value of capacitance in any given capacitor is found from:

$$C = \frac{0.0885 \times K \times A \times (N - 1)}{T}$$

Where: C = capacitance in picofarads (pF)
   K = dielectric constant
   A = area of one of the plates (L × W), assuming that the two plates are identical
   N = number of identical plates
   T = thickness of the dielectric

Breakdown voltage

The capacitor works by supporting an electrical field between two metal plates. This potential, however, can get too large. When the electrical potential gets too large, free electrons in the dielectric material (there are a few, but not many, in any insulator) may flow. If a stream of electrons gets started, then the dielectric could break down and allow a current to pass between the plates. The capacitor is then said to be *shorted*.

The maximum breakdown voltage of the capacitor must not be exceeded. However, for practical purposes there is a smaller voltage called the *dc working voltage* (WVdc) rating that defines the maximum safe voltage that can be applied to the capacitor. Typical values found in common electronic circuits are from 8 to 1,000 WVdc.

Circuit symbols for capacitors

The circuit symbols used to determine fixed-value capacitors are shown in Fig. 7-10A, and for variable capacitors in Fig. 7-10B. Both symbols are common. In certain types of capacitor, the curved plate shown on the left in Fig. 7-10A is usually the outer plate—the one closest to the outside package of the capacitor. This end of the capacitor is often indicated with a color band next to the lead attached to that plate.

The symbols for the variable capacitor are shown in Fig. 7-10B. This symbol is the fixed-value symbol with an arrow through the plates. Small trimmer and padder capacitors are often denoted by the symbol of Fig. 7-10C. The variable set of plates is designated by the arrow.

Fixed capacitors

There are several types of fixed capacitor found in typical electronic circuits. These are classified by dielectric type: paper, Mylar, ceramic, mica, polyester, and others.

The construction of old-fashioned paper capacitors is shown in Fig. 7-11. They consist of two strips of metal foil sandwiched on both sides of a strip of paraffin wax paper. The strip sandwich is then rolled up into a tight cylinder. This rolled up cylinder is then packaged in either a hard plastic, bakelite, or paper-and-wax case. When the case is cracked or the wax end plugs are loose, replace the capacitor even though it tests good—it won't be for long. Paper capacitors come in values from about 300 pF to about 4 μF. The breakdown voltages will be 100 to 600 WVdc.

The paper capacitor is used for a number of different applications in older circuits, such as bypassing, coupling, and dc blocking. Unfortunately, no component is perfect. The long rolls of foil used in the paper capacitor exhibit a significant amount of stray inductance. As a result, the paper capacitor is not used for high frequencies. Although they are found in some shortwave receiver circuits, they are rarely or never used at VHF.

In modern applications, or when servicing older equipment that used paper capacitors, use a Mylar dielectric capacitor in place of the paper capacitor. Select a unit with exactly the same capacitance rating, and a WVdc rating that is equal to or greater than the original WVdc rating.

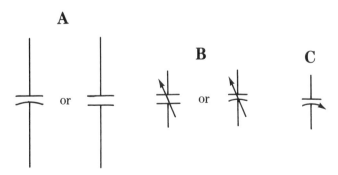

7-10   Capacitor circuit symbols: (A) fixed capacitors, (B) variable capacitors, (C) alternate variable capacitor symbol often used for trimmer/padder capacitors.

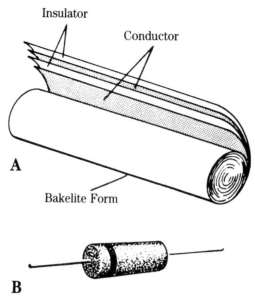

7-11   Construction of a paper capacitor.

Several different forms of ceramic capacitors are shown in Fig. 7-12. These capacitors come in value from a few picofarads up to 0.5 μF. The working voltages range from 400 WVdc to more than 30,000 WVdc. Common disk ceramic capacitors are usually rated at either 600 WVdc or 1,000 WVdc. The tubular ceramic capacitors are typically much smaller in value than disk or flat capacitors, and are used extensively in VHF and UHF circuits for blocking, decoupling, bypassing, coupling, and tuning.

The feedthrough type of ceramic capacitor is used to pass dc and low-frequency ac lines through a shielded panel. These capacitors are often used to filter or decouple lines that run between circuits that are separated by the shield for purposes of electromagnetic interference (EMI) reduction.

Ceramic capacitors are often rated by *temperature coefficient*. This specification is the change of capacitance per change of temperature in degrees Celsius. A "P" prefix indicates a positive-temperature coefficient, an "N" indicates a negative-temperature coefficient, and the letters "NPO" indicate a zero-temperature coefficient (NPO stands for *negative positive zero*).

Do not guess at these ratings when servicing a piece of electronic equipment. Use exactly the same temperature coefficient as the original manufacturer used. Nonzero-temperature coefficients are often used in oscillator circuits to temperature compensate the oscillator's frequency drift.

Several different types of mica capacitor are shown in Fig. 7-13. The *fixed-mica* capacitor consists of either metal plates on either side of a sheet of mica, or a sheet

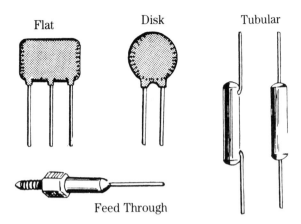

7-12 Forms of ceramic capacitors.

of mica that is silvered with a deposit of metal on either side. The range of values for mica capacitors tends to be 50 pF to 0.02 μF at voltages in the range of 400 WVdc to 1,000 WVdc.

The mica capacitor shown in Fig. 7-13B is called a *silvered mica* capacitor. These capacitors are low-temperature coefficients, although for most applications an NPO disk ceramic will service better than all but the best silvered mica units. Mica capacitors are typically used for tuning and other uses in higher-frequency applications.

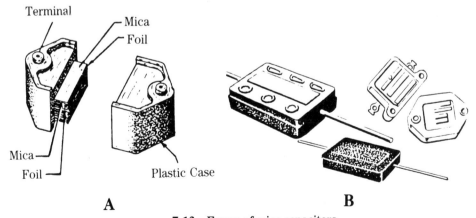

7-13 Forms of mica capacitors.

Other capacitors

Today the equipment designer has a number of different dielectric capacitors available that were not commonly available (or not available at all) a few years ago. The polycarbonate, polyester, and polyethylene capacitors are used in a wide variety of applications where the capacitors just discussed once ruled supreme. In digital circuits we find tiny 100 WVdc capacitors carrying ratings of 0.01 μF to 0.1 μF. These are used for decoupling the noise on the +5 Vdc power supply line.

In circuits such as timers and op amp Miller integrators, where the leakage resistance across the capacitor becomes terribly important, we might want to use a polyethylene capacitor. Check current catalogs for various old- and new-style capacitors—the applications paragraph in the catalog will tell you which applications they can be used in, which is a guide to the type of antique capacitor it will replace.

Variable capacitors

Like all capacitors, variable capacitors are made by placing two sets of metal plates parallel to each other (Fig. 7-14A), separated by a dielectric of air, mica, ceramic, or

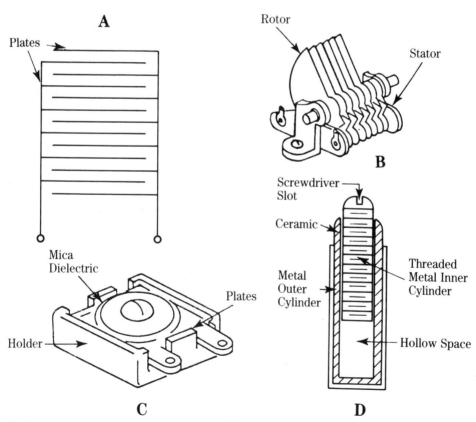

7-14 Variable capacitors: (A) basic form, (B) typical air dielectric variable capacitor, (C) mica compression capacitor, (D) piston capacitor.

a vacuum. The difference between variable and fixed capacitors is that, in variable capacitors, the plates are constructed in such a way that the capacitance can be changed. There are two principal ways to vary the capacitance: either the spacing between the plates is varied, or the cross-sectional area of the plates that face each other is varied.

Figure 7-14B shows the construction of a typical variable capacitor used for the main tuning control in radio receivers. The capacitor consists of two sets of parallel plates. The *stator* plates are fixed in their position, and are attached to the frame of the capacitor. The *rotor* plates are attached to the shaft that is used to adjust the capacitance.

Another form of variable capacitor found in radio receivers is the *compression capacitor* shown in Fig. 7-14C. It consists of metal plates separated by sheets of mica dielectric. In order to increase the capacitance, the manufacturer may increase the area of the plates and mica, or the number of layers (alternating mica/metal) in the assembly. The entire capacitor will be mounted on a ceramic or other form of holder. If mounting screws or holes are provided, then they will be part of the holder assembly.

Still another form of variable capacitor is the *piston capacitor* shown in Fig. 7-14D. This type of capacitor consists of an inner cylinder of metal coaxial to, and inside of, an outer cylinder of metal. An air, vacuum, or (as shown) ceramic dielectric separates the two cylinders. The capacitance is increased by inserting the inner cylinder further into the outer cylinder.

The small compression or piston-style variable capacitors are sometimes combined with air-variable capacitors. The smaller capacitor used in conjunction with the larger air variable is called a *trimmer capacitor*. These capacitors are often mounted directly on the air-variable frame (Fig. 7-15A), or very close by in the circuit. In many radios, the "trimmer" is actually part of the air-variable capacitor.

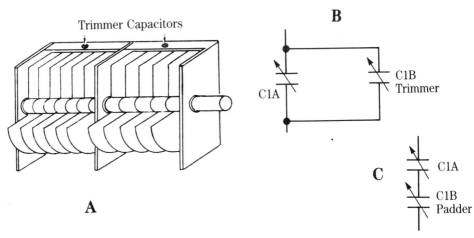

7-15 Use of trimmer and padder capacitors: (A) Small variable capacitors mounted on side of an air variable. (B) A trimmer capacitor is wired in parallel with the main capacitor. (C) A padder capacitor is wired in series with the main capacitor.

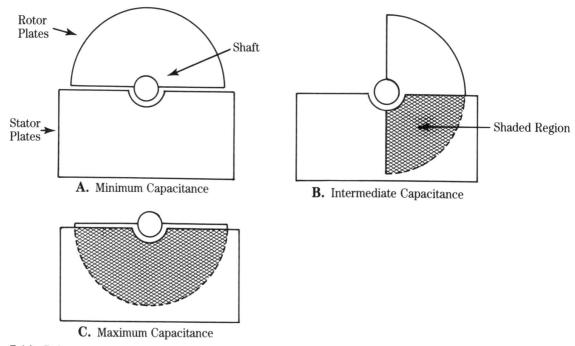

Rotor Plates

Shaft

Stator Plates

A. Minimum Capacitance

Shaded Region

B. Intermediate Capacitance

C. Maximum Capacitance

7-16 Relative position of stator and rotor plates determines capacitance of the air variable. (A) Fully unmeshed plates indicate minimum capacitance. (B) Partially meshed plates are an intermediate capacitance. (C) Fully meshed indicated maximum capacitance.

There are actually two uses for small variable capacitors in conjunction with the main tuning capacitor in radios. First, there is the true "trimmer," a small-valued variable capacitor in parallel with the main capacitor (Fig. 7-15B). These capacitors are used to trim the exact value of the main capacitor. The other form of small capacitor is the *padder* capacitor (Fig. 7-15C), which is connected in series with the main capacitor. Only the parallel connected capacitor is properly called a trimmer.

Air-variable main tuning capacitors   The capacitance of an air-variable capacitor at any given setting is a function of how much of the rotor plate set is shaded by the stator plates. In Fig. 7-16A, the rotor plates are completely outside of the stator plate area. Because the shading is zero, the capacitance is minimum.

In Fig. 7-16B, however, the rotor-plate set has been slightly meshed with the stator plate, so some of its area is shaded by the stator. The capacitance in this position is at an intermediate value. Finally, in Fig. 7-16C the rotor is completely meshed with the stator so the cross-sectional area of the rotor that is shaded by the stator is maximum. Therefore, the capacitance is also maximum. Remember these two rules:

1. Minimum capacitance is found when the rotor plates are completely un-meshed with the stator plates.

2. Maximum capacitance is found when the rotor plates are completely meshed with the stator plates.

7-17 Single-section air variable.

Figure 7-17 shows a typical single-section variable capacitor. The stator plates are attached to the frame of the capacitor, which in most radio circuits is grounded. Front and rear plates have bearing surfaces to ease the rotor's action. These capacitors were often used in early multi-tuning knob TRF radio receivers (the kind where each RF tuned circuit had it's own selector knob). But since that design was not very good, the *ganged variable capacitor* (Fig. 7-18) was invented. These capacitors are basically two or, in the case of Fig. 7-18, three variable capacitors mechanically ganged on the same rotor shaft.

In Fig. 7-18 all three sections of the variable capacitor have the same capacitance, so they are identical to each other. If this capacitor is used in a superheterodyne radio, the section used for the local oscillator tuning must be "padded" with a series capacitance in order to reduce the overall capacitance. This is done to permit the higher frequency LO to track with the RF amplifiers on the dial.

In many superheterodyne radios you will find variable tuning-capacitors in which one section (usually the front section) has fewer plates than the RF amplifier

7-18 Three-section "ganged" air variable.

7-19   Cut-plate capacitor uses a smaller capacitor for the LO of a superheterodyne to ensure tracking with the larger RF-section capacitor.

section (Fig. 7-19). These capacitors are sometimes called *cut-plate capacitors* because the LO section plates are cut to permit tracking of the LO with the RF.

Straight-line capacitance vs. straight-line frequency capacitors   The variable capacitor shown in Fig. 7-16 has the rotor shaft in the geometric center of the rotor plate half-circle. The capacitance of this type of variable capacitor varies directly with the rotor shaft angle. As a result, this type of capacitor is called a *straight-line capacitance* model. Unfortunately, as you will see in a later section, the frequency of a tuned circuit based on inductors and capacitors is not a linear function of capacitance.

If a straight-line capacitance unit is used for the tuner, then the frequency units on the dial will be cramped at one end and spread out at the other (you've probably seen such radios). But some capacitors have an offset rotor shaft (Fig. 7-20) that compensates for the nonlinearity of the tuning circuit. The shape of the plates, and the location of the rotor shaft, are designed to produce a linear relationship between the shaft angle and the resonant frequency of the tuned circuit in which the capacitor is used.

7-20   Off-center rotor shaft and custom-shaped plates make this capacitor straight-line wavelength (or frequency), instead of straight-line capacitance.

Variable capacitor cleaning note   Antique radio buffs often find that the main tuning capacitors in their radios are full of crud, grease and dust. Likewise, ham radio operators working the hamfest circuit looking for linear amplifier and antenna tuner parts often find just what they need, but the thing also has scum, crud, grease and other stuff.

There are several things that can be done about it. First, try using dry compressed air. It will remove dust, but not grease. Aerosol cans of compressed air can be bought from many sources, including automobile parts stores and photography stores.

Another method, if you have the hardware, is to clean the capacitor ultrasonically. The ultrasonic cleaner, however, is expensive, so unless you already have one don't rush out to buy one.

Still another way is to use a product such as *Birchwood Casey Gun Scrubber.* This product is used to clean firearms, and is available in most gun shops. Firearms become gooped up because gun grease, oil, unburned powder, and burned powder residue combine to create a crusty mess that's every bit as hard to remove as capacitor gunk. A related product is the degunking compound used by auto mechanics.

At one time, carbon tetrachloride was used for this purpose, and you will see it listed in old radio books. However, "carbon tet" is now well recognized as a health hazard. *Do not use carbon tetrachloride for cleaning,* despite advice to the contrary found in old radio books.

Capacitors in ac circuits

When an electrical potential is applied across a capacitor, current will flow, as charge is stored in the capacitor. As the charge in the capacitor increases, the voltage across the capacitor plates rises until it equals the applied potential. At this point the capacitor is fully charged, and no further current will flow.

Figure 7-21 shows an analogy for the capacitor in an ac circuit. The actual circuit is shown in Fig. 7-12A, and consists of an ac source connected in parallel across the capacitor (C).

The mechanical analogy is shown in Fig. 7-12B. The "capacitor" (C) consists of a two-chamber cylinder in which the upper and lower chambers are separated by a flexible membrane or diaphragm. The "wires" are pipes to the "ac source," which is a pump. As the pump moves up and down, pressure is applied to first one side of the diaphragm then the other, alternately forcing fluid to flow into and out of the two chambers of the "capacitor."

This ac circuit mechanical analogy is not perfect, but it works for our purposes. Now let's apply these ideas to the electrical case. In Fig. 7-22 we see a capacitor connected across an ac sinewave source. In Fig. 7-22A, the ac source is positive, so negatively charged electrons are attracted from plate A to the ac source, and electrons from the negative terminal of the source are repelled towards plate B of the capacitor.

On the alternate half-cycle (Fig. 7-22B), the polarity is reversed, so electrons from the new negative pole of the source are repelled toward plate A of the

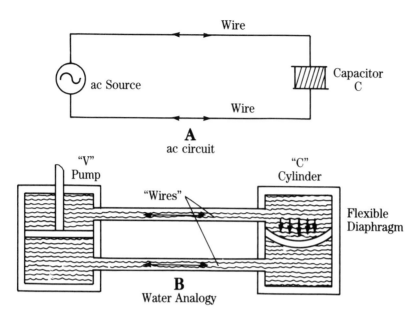

7-21 (A) Capacitor in ac circuit. (B) Water circuit analogy for capacitor in an ac circuit.

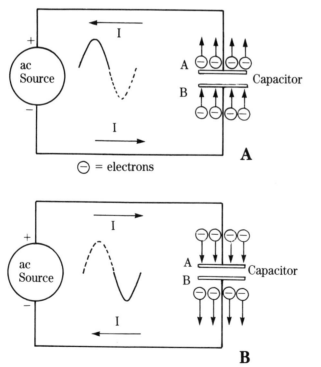

7-22 Action of the capacitor in an ac circuit: (A) When plate A is positive with respect to plate B, electrons rush into plate B and away from plate A. (B) On the alternate half-cycle the opposite action occurs.

capacitor, and electrons from plate B are attracted towards the source. Thus, current will flow in and out of the capacitor on alternating half-cycles of the ac source.

Voltage and current in capacitor circuits

Consider the circuit in Fig. 7-23: an ac source (V) is connected in parallel with the capacitor (C). It is the nature of a capacitor to oppose these changes in the applied voltage, the inverse of the action of an inductor. As a result, the voltage (V) lags behind the current (I) by 90 degrees. These relationships are shown in terms of sinewaves in Fig. 7-23B, and in vector form in Fig. 7-23C.

Do you want to remember the difference between the action of inductors (L) and capacitors (C) on the voltage and current? In earlier texts, they used the letter E to denote voltage, so it made a little mnemonic:

ELI THE ICE MAN

"ELI the ICE man" suggests that in the inductive (L) circuit, the voltage (E) comes before the current (I)—ELI, and in a capacitive (C) circuit the current comes before the voltage—ICE.

The action of a circuit containing a resistance and capacitance is shown in Fig. 7-24A. As in the case of the inductive circuit, there is no phase shift across the resistor, so the **R** vector points in the "east" direction (Fig. 7-24B). The voltage across the capacitor, however, is phase shifted 90 degrees, so its vector points "south." The total resultant phase shift (Δ) is found using the Pythagorean rule to calculate the angle between V_r and V_t.

The impedance of the RC circuit is found in exactly the same manner as the impedance of an RL circuit; that is the root of the sum of the squares:

$$Z = \sqrt{R^2 + (X_c)^2}$$

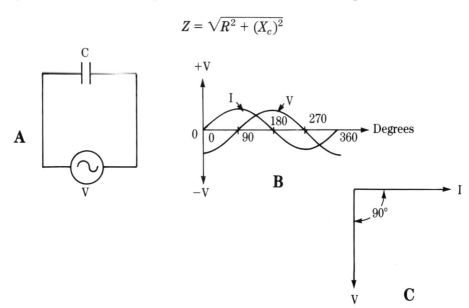

7-23   (A) Capacitor in an ac circuit. (B) Phase relationships. (C) Vector representation.

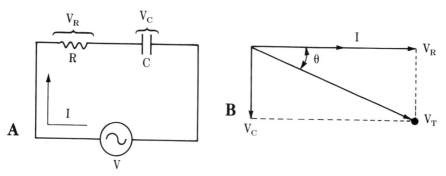

7-24   (A) Resistor-capacitor (RC) circuit. (B) Vector representation.

LC Resonant Circuits

When you use an inductor (L) and a capacitor (C) together in the same circuit, the combination forms an LC resonant circuit, also sometimes called a *tank circuit* or *resonant tank circuit*. These circuits are used to tune a radio receiver. There are two basic forms of LC resonant tank circuit: series (Fig. 7-25A) and parallel (Fig. 7-25B). These circuits have much in common and much that makes them fundamentally different from each other.

The condition of resonance occurs when the capacitive reactance (X_c) and inductive reactance (X_L) are equal. As a result, the resonant tank circuit shows up as purely resistive at the resonant frequency, and as a complex impedance at other frequencies. The LC resonant tank circuit operates by an oscillatory exchange of energy between the magnetic field of the inductor, and the electrostatic field of the capacitor, with a current between them carrying the charge.

Because the two reactances are both frequency dependent, and because they are inverse to each other, the resonance occurs at only one frequency (f_r). We can calculate the standard resonance frequency by setting the two reactances equal to each other and solving for f. The result is:

$$f = \frac{1}{2\pi\sqrt{LC}}$$

Series-resonant circuits

The series-resonant circuit (Fig. 7-26A), like other series circuits, is arranged so that the terminal current (I) from the source (V) flows in both components equally. The vector diagrams of Fig. 7-26B through 7-26D show the situation under three different conditions. In Fig. 7-26B, the inductive reactance is larger than the capacitive reactance, so the excitation frequency is greater than f_r. Note that the voltage drop across the inductor is greater than that across the capacitor, so the total circuit looks like it contains a small inductive reactance.

In Fig. 7-26C, the situation is reversed: the excitation frequency is less than the resonant frequency, so the circuit looks slightly capacitive to the outside world.

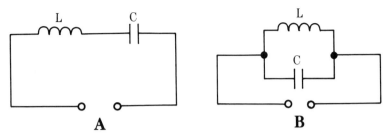

7-25 (A) Series-resonant circuit. (B) Parallel-resonant circuit.

Finally, in Fig. 7-26D the excitation frequency is at the resonant frequency, so $X_c = X_L$ and the voltage drops across the two components are equal but of opposite phase.

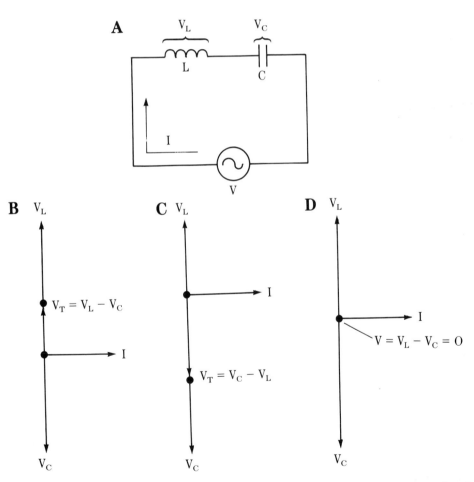

7-26 (A) Voltages and current in a series-resonant circuit. (B) Vector relationship for $X_L > X_c$. (C) *For* $X_c > X_L$. (D) $X_c = X_L$.

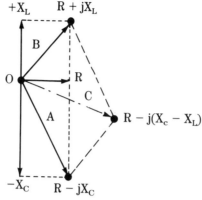

7-27 Vector relationships.

In a circuit that contains a resistance, inductive reactance, and a capacitive reactance, there are three vectors to consider (Fig. 7-27), plus a resultant vector. As in the other circuit, the "north" direction represents X_L, the "south" direction represents X_c, and the "east" direction represents R.

Using the parallelogram method, we first construct a resultant vector for the R and X_c, which is shown as vector **A**. Next, we construct the same kind of vector **B** for R and X_L. The resultant, C, is made using the parallelogram method on **A** and **B**. Vector **C** represents the impedance of the circuit: the magnitude is represented by the length, and the phase angle by the angle between **C** and R.

Figure 7-28A shows a series-resonant LC tank circuit, and Figs. 7-28B and 7-28C shows the current and impedance as a function of frequency. The series-resonant circuit has a low impedance at its resonant frequency and a high impedance at all other frequencies. As a result, the line current (I) from the source is maximum at the resonant frequency and the voltage across the source is minimum.

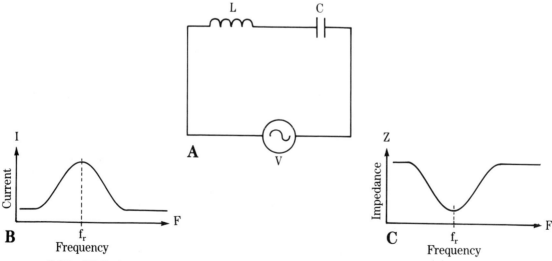

7-28 (A) Series-resonant circuit. (B) Current-vs.-frequency. (C) Impedance-vs.-frequency.

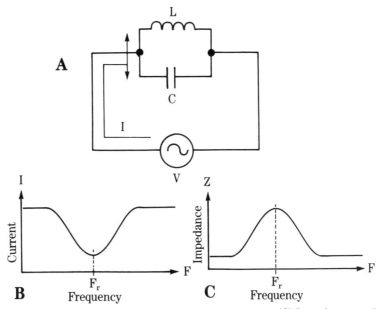

7-29   (A) Parallel-resonant circuit. (B) Current-vs.-frequency. (C) Impedance-vs.-frequency.

Parallel-resonant circuits

The parallel-resonant tank circuit (Fig. 7-29A) is the inverse of the series-resonant circuit. The line current from the source splits and flows to the inductor and capacitor separately. The parallel-resonant circuit has its highest impedance at the resonant frequency, and a low impedance at all other frequencies (Fig. 7-29C). Thus, the line current from the source is minimum at the resonant frequency (Fig. 7-29B), and the voltage across the LC tank circuit is maximum. This fact is important in radio tuning circuits, as you will see in due course.

Tuned RF/IF Transformers

Most resonant circuits used in radio receivers are actually transformers that couple the signal from one stage to another. Figure 7-30 shows several popular forms of tuned, or coupled, RF/IF tank circuits. In Fig. 7-30A, one winding is tuned while the other is untuned. In the configurations shown, the untuned winding is the secondary of the transformer.

    This type of circuit is often used in transistor and other solid-state circuits, or when the transformer has to drive either a crystal or mechanical bandpass filter circuit. In the reverse configuration (L1 = output, L2 = input), the same circuit is used for the antenna coupling network, or as the interstage transformer between RF amplifiers in TRF radios.

    The circuit in Fig. 7-30B is a parallel-resonant LC tank circuit equipped with a low-impedance tap. This type of circuit is used to drive a crystal detector or other low-impedance load. Another circuit for driving a low-impedance load is shown in

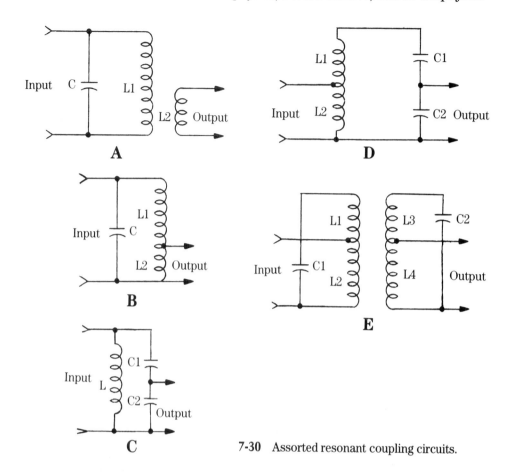

7-30   Assorted resonant coupling circuits.

Fig. 7-30C. This circuit splits the capacitance that resonates the coil into two series capacitors.

As a result, we have a capacitive voltage divider. The circuit in Fig. 7-30D uses a tapped inductor for matching low-impedance sources (e.g., antenna circuits), and a tapped capacitive voltage divider for low-impedance loads. Finally, the circuit in Fig. 7-30E uses a tapped primary and tapped secondary winding in order to match the two low-impedance loads while retaining the sharp bandpass characteristics of the tank circuit.

Construction of RF/IF transformers

Tuned RF/IF transformers built for radio receivers are typically wound on a common cylindrical form, and surrounded by a metal shield can that prevents interaction of the fields of coils that are in close proximity to each other.

Figure 7-31A shows the schematic for a typical RF/IF transformer, while the sectioned view shows one form of construction. This method of building transformers was common at the beginning of World War II, and continued into the early transistor era. The capacitors were built into the base of the transformer, while the

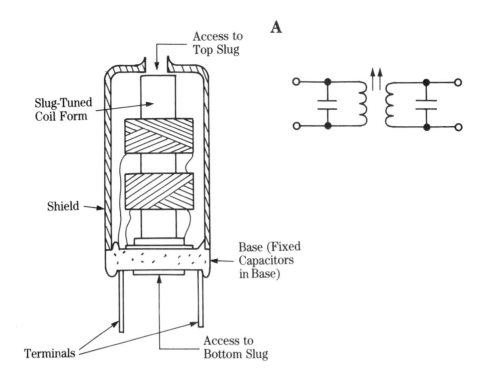

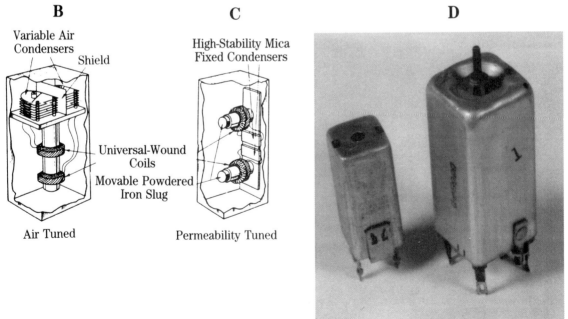

7-31 Construction of tuned transformers: (A) relatively modern form, (B) older form using air variable capacitors, (C) slug-tuned older form, (D) physical form of IF/RF transformer.

tuning slugs were accessed from holes in the top and bottom of the assembly. In general, expect to find the secondary at the bottom hole, and the primary at the top hole.

Two older forms of construction (pre-WWII with some immediately post-WWII) are shown in Figs. 7-31B and 7-31C. In Fig. 7-31B, there are a pair of air-variable capacitors at the top end to resonate the coils. The capacitors are adjusted through holes in the metal shield. An alternative method, using slug tuning, is shown in Fig. 7-31C.

The term *universal wound* refers to a cross-winding system that minimizes the interwinding capacitance of the inductor, and therefore raises the self-resonant frequency of the inductor. Actual examples of RF/IF transformers are shown in Fig. 7-31D. The smaller type tends to be post-WWII, while the larger type is pre-WWII.

Bandwidth of RF/IF transformers

Figure 7-32A shows a parallel resonant RF/IF transformer, while Fig. 7-32B shows the usual construction in which the two coils (L1 and L2) are wound at distance d apart on a common cylindrical form. The *bandwidth* of the RF/IF transformer is the frequency difference between the frequencies, where the signal voltage across the output winding falls off -6 dB from the value at the resonant frequency (f_r), as shown in Fig. 7-32C.

If F_1 and F_2 are -6 dB (also called the -3 dB point, when signal power is measured instead of voltage) frequencies, then the bandwidth (BW) is F_2-F_1. The shape of the frequency-response curve in Fig. 7-32C is said to represent *critical coupling*.

An example of a *subcritical* or *undercoupled* RF/IF transformer is shown in Fig. 7-33. As shown in Figs. 7-33A and 7-33B, the windings are farther apart than in

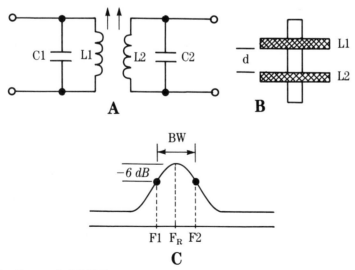

7-32   Critically coupled IF/RF transformer: (A) circuit, (B) placement on coil form, (C) frequency response.

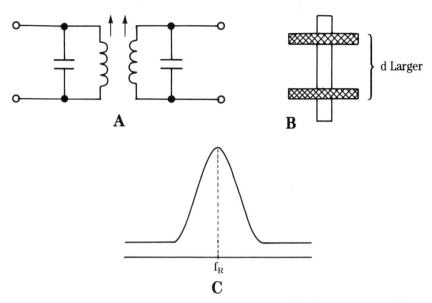

7-33 Subcritically coupled IF/RF transformer: (A) circuit, (B) placement on coil form, (C) frequency response.

the critically coupled case, so the bandwidth (Fig. 7-33C) is much narrower than in the critically coupled case. The subcritically coupled RF/IF transformer is often used in shortwave or communications receivers in order to allow the narrower bandwidth to discriminate against adjacent channel stations.

The *overcritically coupled* RF/IF transformer is shown in Fig. 7-34. Here we note in Figs. 7-34A and 7-34B that the windings are closer together, so the bandwidth (Fig. 7-34C) is much broader. In some radio schematics, service manuals and early

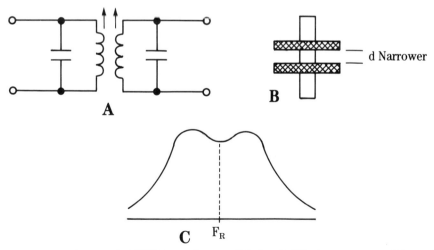

7-34 Overcritically coupled IF/RF transformer: (A) circuit, (B) placement on coil form, (C) frequency response.

textbooks, this form of coupling was sometimes called "high fidelity" coupling because it allowed more of the sidebands of the signal, which carry the audio modulation, to pass with less distortion of frequency response.

The bandwidth of the resonant tank circuit, or the RF/IF transformer, can be summarized in a specification called Q. The Q of the circuit is the ratio of the bandwidth to the resonant frequency: $Q = BW/f_r$. An overcritically coupled circuit has a low Q, while a narrow-bandwidth subcritically coupled circuit has a high Q.

A resistance in the LC tank circuit will cause it to broaden; that is to lower its Q. The *loaded Q*—i.e., Q when a resistance is present, as in Fig. 7-35A—is always less than the unloaded Q. In some radios, a switched resistor (Fig. 7-35B) is used to allow the user to broaden or narrow the bandwidth. This switch might be labeled *fidelity*, or *tone*, or something similar.

RF Amplifier Circuits

The RF amplifier formed the basis of the TRF radio, and is used to preselect stations in better quality superheterodyne radios. Figure 7-36 shows a two-stage cascade-triode RF amplifier, as might be found in a TRF receiver.

Transformer L1 is the antenna tuner; currents flowing in L1A induce currents in secondary winding L1B. Because L1B and C1 form a parallel-resonant LC tank circuit, the voltage appearing across the grid-cathode circuit of V1 is maximum at the resonant frequency and nearly zero at all other frequencies.

As a result, V1 amplifies only the resonant frequency. This amplified signal is

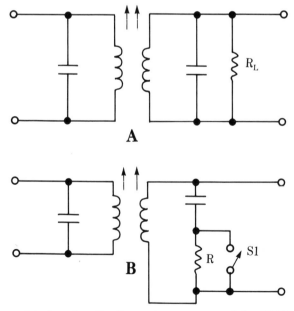

7-35   Resistors used to broaden the frequency response of the IF/RF transformer: (A) parallel version, (B) series version (shown with "fidelity" switch).

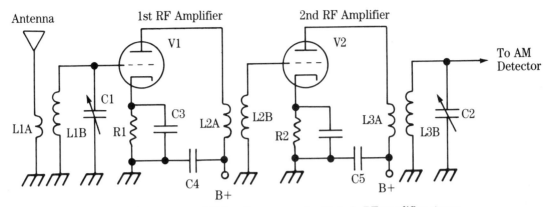

7-36 Two-stage TRF radio uses a pair of triode RF amplifier stages.

passed from the plate of V1 through transformer L2 to the grid of V2 where it receives additional amplification in a similar circuit. The output of the second RF amplifier (secondary of L3) is fed to the AM detector stage.

Unfortunately, there is a major problem with the triode RF-amplifier of Fig. 7-36: it tends to oscillate at or near the resonant frequency. The problem is the interelectrode capacitances of the triode (Fig. 7-37A). There is capacitance between the grid and cathode (C_{gk}), between the grid and plate (C_{gp}), and between the cathode and plate (C_{pk}). Of these capacitances, C_{gp} and C_{gk} are the most important.

Figure 7-37B shows how these capacitances would look if they were discrete capacitors wired into the circuit. The interelectrode capacitance forms a capacitive voltage-divider feedback network that applies a sample of the plate signal to the grid. At some frequency near resonance, the feedback signal is in phase with the grid signal, so the circuit will oscillate. This condition is called *positive feedback*.

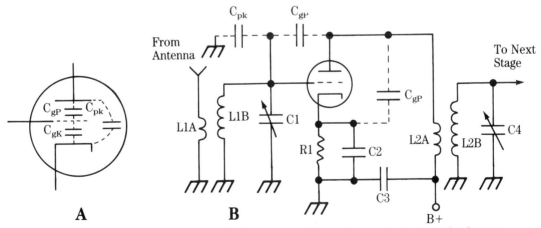

7-37 Source of some trouble: (A) the interelectrode capacitances inside the tube form a feedback network that can cause oscillations, (B) circuit accounting for interelectrode capacitances.

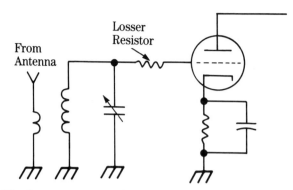

7-38 *Losser resistor* method of reducing tendency to oscillate.

A couple of techniques are used to reduce the tendency to oscillate. One early (and not very effective) method is shown in Fig. 7-38. A series-connected *losser resistor* is wired into the circuit between the LC tank circuit and the grid of the vacuum tube.

A superior method for preventing self-oscillation of the triode amplifier is shown in Fig. 7-39. In this case, a small sample of the output signal is fed back external to the tube. This negative feedback signal passes through a *neutralization capacitor* (C2), and has an amplitude and phase that exactly cancels the internal positive feedback produced by the interelectrode capacitance. The capacitance of C2 is adjusted to a point where the stage is stable (nonoscillating) across the entire tuning range of the RF amplifier.

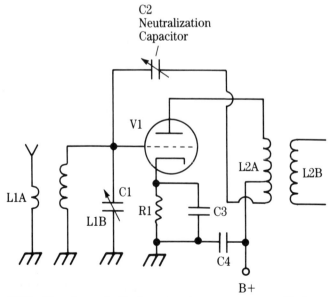

7-39 Use of a neutralization capacitor to prevent oscillation.

The triode RF amplifier is more susceptible to oscillation than either tetrode or pentode RF amplifiers because the interelectrode capacitances are larger. In many tetrode and pentode RF amplifiers no neutralization whatsoever is needed. For that reason, triodes were less popular after the late 1920s when tetrodes became widely available.

Another stability problem seen in RF amplifiers is due to feedback between stages in a cascade chain through the common dc power-supply connection (see again Fig. 7-36). Figure 7-40 shows two *decoupling* schemes used to prevent this problem. In Fig. 7-40A, a *radio frequency choke* (RFC1) is connected in series with the B+ line and the transformer primary.

The RFC offers a high impedance to RF, while maintaining a low dc resistance for the plate current. Capacitor C1 offers a low impedance to RF signals, and so shunts them to ground. Figure 7-40B is similar in function, but replaces the RFC with a series resistor.

Figure 7-41 shows a two-stage RF amplifier, similar to Fig. 7-36, but with the decoupling resistors connected in place, and neutralization provided for each stage. In addition, the three main tuning capacitors (C2A, C3A, and C4A) are mechanically ganged together to permit single-knob tuning of the radio. Shunted across each main tuning capacitor section is a smaller-value trimmer capacitor.

These trimmers are used to align the three tuning circuits together so that they will track each other across the band. The variable capacitor in the antenna circuit is called the *antenna trimmer*. It is used to tune L1A to the same frequency that the radio receives.

Superheterodyne Receivers

Recall from chapter 1 that the superheterodyne radio converts the RF signal frequency down to an intermediate frequency where most of the gain and selectivity

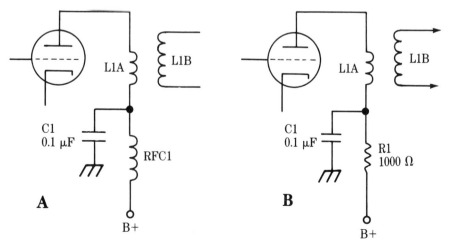

7-40 Prevention of oscillation between stages requires a decoupling network: (A) RF choke-type circuit, (B) resistor-type circuit.

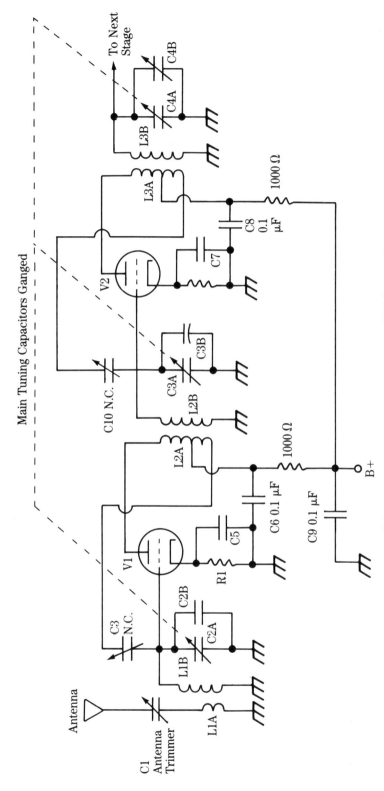

7-41 Complete circuit for a two-stage TRF front-end.

is provided. The frequency conversion comes from heterodyning ("beating") the RF signal against a local oscillator signal in a mixer circuit, and then tuning the output of the mixer to the difference frequency (RF-LO or LO-RF) *Note* : The sum frequency can also be used, but rarely was in older radios.

Figure 7-42 shows a partial circuit of a triode-based superhet receiver. The RF amplifier is nearly identical to the RF amplifier from the TRF radio shown earlier (Figs. 7-36 and 7-41). The RF amplifier is not strictly necessary for superhet operation, but all good quality receivers use one.

The purpose of the RF amplifier is to provide a small amount of gain, but a large amount of *preselection*. Recall that preselection improves the image response of the radio—a problem unique to superhets.

The output of the RF amplifier tube (V1) is coupled to the mixer stage (V2) through transformer L2. Inductor L2B and C2 form a resonant circuit at the same frequency as the input tuned circuit, so it develops a stronger signal at the grid of the mixer. The LO signal is applied to the circuit through the cathode, although other routes are possible. In order to accommodate single-knob tuning, the two RF amplifier capacitors and the LO capacitor are mechanically ganged to the same shaft, as indicated by the dotted line.

The LO signal has a sufficient amplitude to drive the mixer into nonlinearity, a necessary condition for proper mixing. The difference frequency (IF) is selected by a tuned transformer (T1), which is resonant at the difference frequency, LO-RF or RF-LO.

Superheterodyne local-oscillator circuits

The local oscillator signal is generated in the radio for use in the mixer circuit. There are three basic forms of oscillator circuit used in radios for the LO: Armstrong, Colpitts, and Hartley. Other forms are used occasionally, but these three are the most common. The purpose of the oscillator circuit is to encourage and optimize the very attribute that we try to suppress in RF amplifiers—oscillation.

Barkhausen's Criteria determines whether or not the circuit will oscillate. At the desired frequency: the feedback signal must be in phase (360 degrees) with the input signal, and the total circuit gain must be greater than one (overcoming losses in the bargain). The three different oscillators differ principally in how they achieve the feedback signal.

In Fig. 7-43A we see the Armstrong oscillator circuit, invented by Edwin Armstrong. This circuit uses a small "ticker coil" (L2) that is magnetically coupled to the tuning coil (L1). The Colpitts oscillator (Fig. 7-43B) uses a capacitive voltage divider to create the feedback. In the example shown, the feedback network also resonates the inductor, but this is not always the case. Finally, the Hartley oscillator (Fig. 7-43C) uses a tapped inductor to provide the feedback signal.

Figure 7-44A shows an Armstrong local oscillator feeding a signal to the mixer circuit. While the previous example showed cathode excitation of the mixer stage, this circuit uses grid excitation. Whichever method is used, however, the idea is to use the LO signal to drive the mixer into a nonlinear region to facilitate heterodyning. Other coupling methods and LO output methods are also known, but are too numerous to cover in detail here.

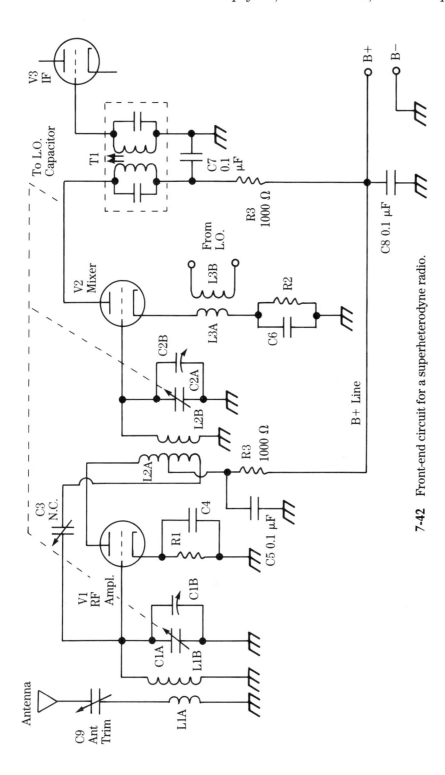

7-42   Front-end circuit for a superheterodyne radio.

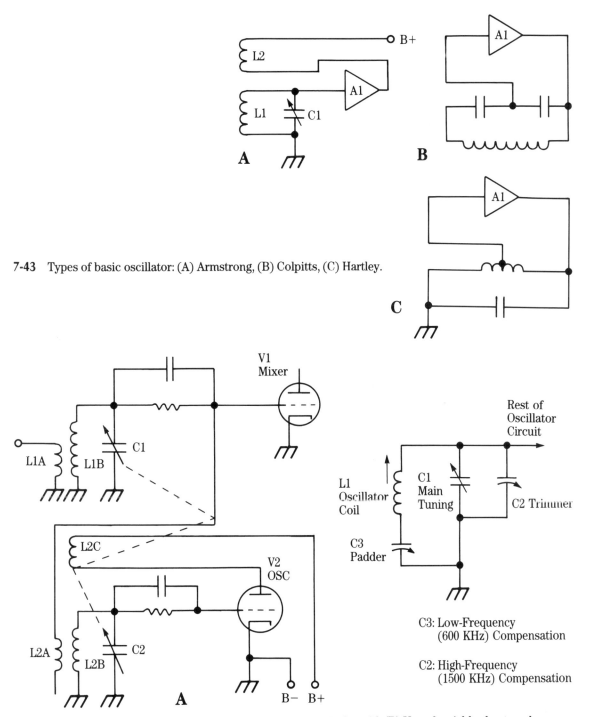

7-43 Types of basic oscillator: (A) Armstrong, (B) Colpitts, (C) Hartley.

7-44 (A) Coupling Armstrong oscillator to mixer via the grid. (B) Use of variable slug-tuned coil, trimmer capacitor, and a padder capacitance allows adjustment of the local oscillator to track with the RF tuned circuits.

If the LO tuning capacitor has the same range of values as the RF tuning capacitor and the mixer grid-tuning capacitor, then there must be some means for permitting the LO to track the RF (the LO and RF are on different frequencies). Unless a special capacitor is used, a circuit such as Fig. 7-44B is needed. In this case, a trimmer (C2) is connected in parallel with the main tuning capacitor (C1), and a padder (C3) connected in series with the inductor.

In general, the trimmer is used to adjust the high-frequency end of the dial, while the inductor and padder capacitor are used to track the low-frequency end of the dial. For this reason, the padder capacitor is sometimes called the *600 padder*, after the normal 600 KHz adjustment point.

Pentagrid converter circuits

The more modern superhet receivers used a special vacuum tube, such as a 6BE6, that combined the local oscillator and mixer stages into a single tube. The LO and mixer elements shared common main anode and cathode, but different grids. This type of circuit is called a *converter* stage, or a *pentagrid converter* because the tube contains five grids (Fig. 7-45). G1 is used as the LO grid, G2 and G4 are connected

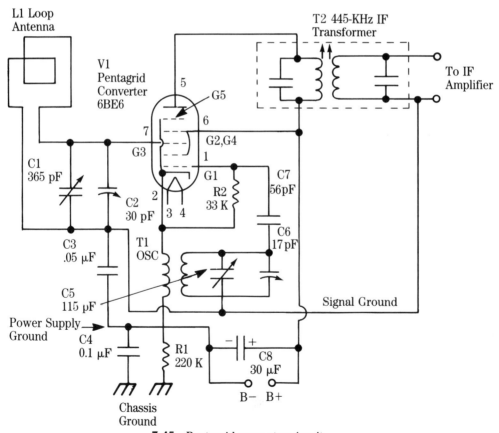

7-45   Pentagrid converter circuit.

together to form the LO plate and mixer screen grids, G3 is the mixer stage RF input, and G5 is a suppressor grid for the mixer function. The LO coil (T1) and tuning capacitor (C5) form the LO resonant circuit, which is connected as a triode oscillator between K, G1 and G2. The RF signal is applied to G3, and the IF is taken from the plate.

The IF amplifier in superhet receivers is basically an RF-amplifier type of circuit, operated at the IF frequency; typically 450 to 460 KHz in older radios, but now standardized at 455 KHz. In most cases, the tuned transformers in the IF amplifier stage are double-tuned, such as T2 in Fig. 7-45.

8

AM Detector and
Automatic Volume-
Control Circuits

THE JOB OF THE AM RADIO DETECTOR CIRCUIT SOMETIMES CALLED *SECOND DETECTOR* or *demodulator*) is to demodulate the amplitude-modulated RF carrier signal, and then recover the modulating audio signal and pass it on to the audio-frequency amplifier stages of the receiver.

The job of the automatic volume control (AVC) circuit, also sometimes called *automatic gain control* (AGC), is to maintain a relatively constant signal level at the output of the radio, even though radio signals being received have differing strengths at the antenna terminals. In this chapter we will study both the detector circuits and the AVC/AGC circuits found in radios.

Review of Amplitude Modulation

In order to frame the discussion of demodulating the AM signal, let's first review the process of amplitude modulation to see what we are dealing with. Figure 8-1A shows an audio tone that we want to send out over an AM transmitter.

For simplicity we will consider only the sinewave case, but the process for speech is the same. The sinewave is symmetrical about the 0 volts axis, and will have a frequency between 20 Hz and 20,000 Hz.

The transmitter radio frequency signal is also a sinewave, but it has a frequency that is very much larger than the audio frequency. Called the *carrier*, the RF signal will have a frequency somewhere between 20 KHz and daylight; for the AM broadcast band the RF carrier will have a frequency between 550 KHz and 1600 KHz.

The process of amplitude modulation causes the low-frequency audio-modulating signal to be superimposed onto the high-frequency carrier (see Fig. 8-1B). The amplitude modulated RF carrier is transmitted by the station and received at the radio antenna to produce a signal that looks very much like Fig. 8-1B.

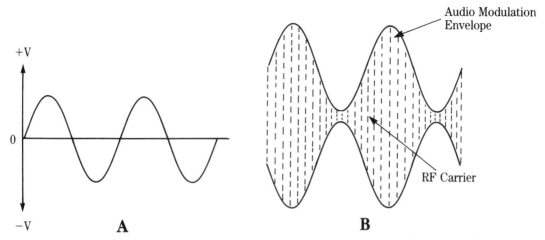

8-1   (A) Audio modulating sinewave signal. (B) Amplitude modulated (AM) carrier signal.

The amplitude of the peaks and valleys in the modulated waveform depends upon the percentage of modulation provided by the transmitter. Note: 100 percent modulation causes the two valleys on each cycle to barely touch the zero baseline. It is the highest percentage of modulation that will not result in distortion of the recovered waveform.

AM detectors

The process of demodulation, or *detection* as it is usually called, is basically the process of removing the RF component and recovering the modulating audio signal undistorted. The job of the demodulator is to offer a nonlinear situation in which the circuit impedance changes with cyclical excursions of the input signal. Because the modulating signal rides on the envelope of the RF carrier, AM demodulator circuits are usually called *envelope detectors*.

Fleming and DeForest detectors

Envelope detection was the earlier job given to the then-new vacuum tubes of J. Ambrose Fleming. The Fleming Valve was a diode tube that could be used as a rectifier, producing unidirectional flow of current. When the anode is positive with respect to the cathode, current will flow through the detector. But when the anode is negative with respect to the cathode, the device is cut off and no current can flow. These effects were discussed in detail in chapter 3 and elsewhere.

Figure 8-2A shows the simplest form of AM detector based on either the diode vacuum tube or a crystal, while Fig. 8-2B shows the waveforms. Today the use of galena has been abandoned in favor of germanium and silicon pn-junction diodes, such as the 1N34, 1N60, ECG-109, or 1N4148. In the example of Fig. 8-2 we see an elementary crystal set type of circuit, but the same circuit can be used at the output of the RF chain of a TRF radio or superheterodyne radio as well.

The waveforms in Fig. 8-2B show the RF carrier and its modulation envelope after the signal has passed through the half-wave rectifier (V1 or D1). There are no

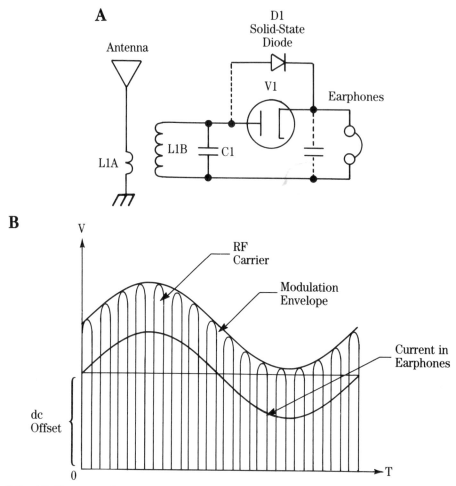

8-2 (A) Vacuum tube or solid-state diode crystal set. (B) Waveform showing detected carrier wave and current in headphones winding.

negative peaks on the carrier because of the rectification action; only positive pulses are present in the circuit. The amplitudes of the pulses follow the modulation envelope of the applied signal.

The effect of the earphones (and the capacitor that is sometimes seen), which form the load for the rectifier is to smooth the waveform by filling in the valleys between RF pulses. The result is a lower-amplitude continuous audio signal in the earphones that is perceived as sound by the listener. Note that the audio signal is superimposed on a small dc offset voltage.

Plate detection

When the triode vacuum tube was invented by Lee DeForest it was placed in service as a more sensitive AM detector. Figure 8-3 shows the basic triode plate detector circuit used in certain very early radio receiving sets. The plate current of vacuum

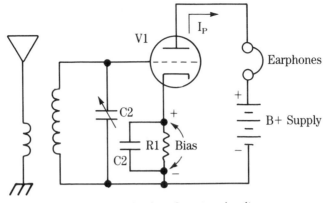

8-3 Triode plate detector circuit.

tube V1 also flows in the cathode circuit, so it will produce a voltage drop across cathode resistor R1.

This bias voltage is set to a point where the tube is almost in cutoff, and is driven into conduction by the applied RF signal. Like the half-wave rectifier, the result of this level of bias is RF pulses in the plate circuit that vary in amplitude in step with the modulation envelope.

When radios progressed sufficiently to allow the use of audio-frequency power amplifiers to drive loudspeakers or earphones, the same type of detector circuit was often used to produce the demodulated audio signal. Figure 8-4A shows a modified form of plate detector that produces a signal suitable for driving a following AF power amplifier output stage. Perhaps this description of this circuit's operation

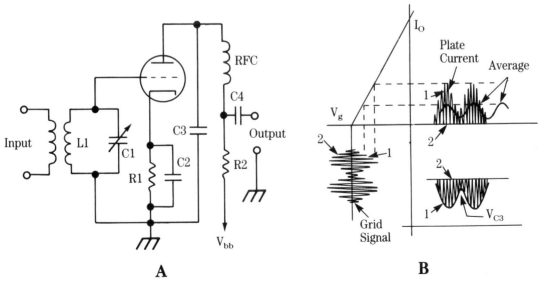

8-4 Plate detection: (A) circuit, (B) waveform relationships.

from an out-of-date Navy radio training manual is the best:

> In a plate detector, the RF signal is first amplified in the plate circuit and then it is detected in the same circuit. . . . A plate detector circuit is shown in Fig. 8-4A. The cathode-bias resistor, R1, is chosen so that the grid bias is approximately at cutoff during the time that an input signal of appropriate strength is applied. Plate current then flows only on the positive swings of grid voltage, during which time average plate current increases. The peak value of the ac input signal is limited to slightly less than the cutoff bias to prevent driving the grid positive on the positive half-cycles of the input signal. Thus, no grid current flows at any time in the input cycle, and the detector does not load the input tuned circuit L1C1.
>
> Cathode bypass capacitor C2 is large enough to hold the voltage across R steady at the lowest audio frequency to be detected in the plate circuit. C3 is the demodulation capacitor across which the AF component is to be developed. R2 is the plate load resistor. The RF choke blocks the RF component from the output [circuit]. R2–C3 has a long time constant with respect to the time for the RF cycle, so that C3 resists any voltage change which occurs at the RF rate. R2–C3 has a short time constant with respect to the time for one AF cycle so the capacitor is capable of charging and discharging at the audio rate.
>
> The action of the plate detector may be demonstrated by the use of the i_p-v_g curve in Fig. 8-4B. On the positive half-cycle of RF input signal (point 1) the plate voltage falls below the V_{bb} supply because of the increased voltage drop across R2 and the RF choke. Capacitor C3 discharges. The discharge current flows clockwise through the circuit, including the tube and C3. The drop across R2 and the RF choke is limited, and the decrease in plate voltage is slight.
>
> On the negative half-cycle of RF input signal (point 2) plate current is cut off and plate voltage rises. Capacitor C3 charges. The charging current flows clockwise around the circuit including the RF choke, R2 and the V_{bb} supply. The voltage drop across R2 and the RF choke, caused by the charging current of C3, checks the rise in plate voltage. Thus, C3 resists voltage changes at the RF frequency. Because C3-R2 has a short time constant with respect to the lowest [frequency] AF signal, the voltage across C3 varies at the AF rate.
>
> The plate detector has excellent selectivity. Its sensitivity (ratio of AF output to RF input) is also greater than that of the diode detector [or the simple plate detector of Fig. 8-3]. However, it is inferior to the diode detector in that it is unable to handle strong signals without overloading. Another disadvantage is that the operating bias will vary with the strength of the incoming signal and thus cause distortion unless a means is provided to maintain the signal input at a constant level. Thus, [in this type of radio] the automatic volume control of manual RF gain-control circuits usually precedes the detector.

Grid-leak detectors

The *grid-leak detector* was invented early in radio history, and became the standard in the late 1920s and early 1930s. But before we discuss this form of detector we need to discuss grid-leak bias.

The grid of the vacuum tube must be placed at a slight negative bias with respect to the cathode. This "C" bias voltage can be supplied by either a battery or negative voltage power supply. Such a bias arrangement is called *fixed bias*.

Alternatively, a bias voltage is provided by inserting a resistor in series with the cathode circuit. The cathode-plate current flowing in that resistor causes a slight voltage drop that places the cathode at a small positive potential above ground potential (0 volts).

Because the grid is fed by a tuned circuit or coil, or has a resistance to ground with no current flowing in it, the grid is essentially at zero volts ground potential. Thus, the positive cathode voltage has the effect of making the grid more negative than the cathode, and accomplishes the bias function. These methods are used extensively in radio circuits to produce bias.

Grid leak bias is little. Figure 8-5A shows a triode amplifier circuit. Normally, the negative bias on the grid retards the flow of electrons from the cathode to the plate.

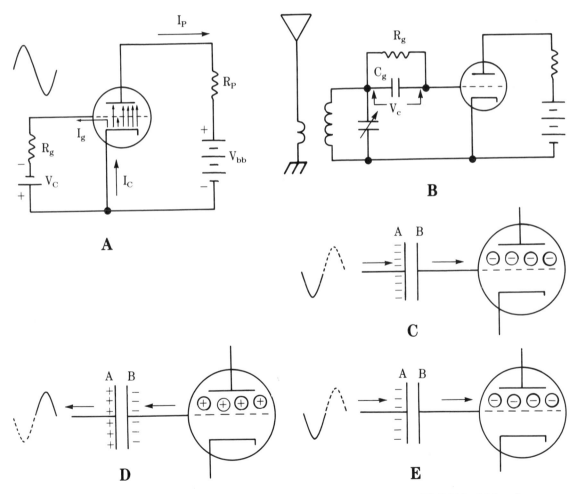

8-5 (A) Some grid current flowing from main current stream in tube. (B) Grid-leak biased circuit. (C) Electron action on negative half-cycle of input signal. (D) Action on positive half-cycle. (E) Action on next negative half-cycle.

Depending upon the value of any fixed bias V_g that might be used, the grid will occasionally draw current on positive excursions of the input signal voltage.

In a grid-leak bias current (Fig. 8-5B) a capacitor is placed in series with the grid circuit. Ignoring the resistor (R_g) shunted across the capacitor for the time being, let's figure out how this circuit works.

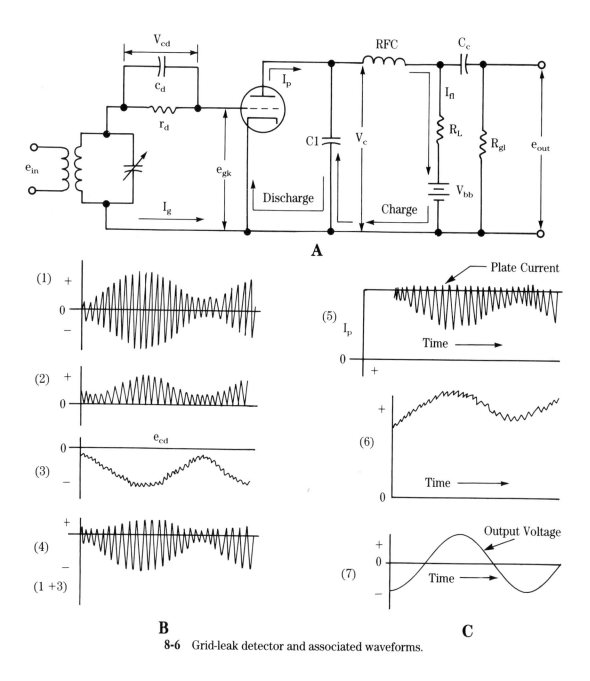

8-6　Grid-leak detector and associated waveforms.

Consider Fig. 8-5C. When the input signal is on its negative half-cycle, electrons are forced away from the negative signal potential onto plate A of the capacitor. Because plate A is now negative, it repels electrons on plate B, which had previously accumulated from grid current and free electrons in the capacitor plate itself.

These plate B electrons are forced onto the grid of the tube causing it to take on a negative charge. When the signal polarity enters the positive excursion (Fig. 8-5D), the electrons from the grid are attracted back to plate B of the capacitor, and the grid now looks positive. On the next negative input-signal excursion the electrons are again forced from plate B of the capacitor onto the tube grid, forcing the grid to become negative.

As the electrons accumulate in the capacitor they form a negative dc voltage (V_c) that biases the grid by an amount proportional to the signal strength. But this process becomes a problem if allowed to continue unabated.

Eventually, the charge on the capacitor will build up to a point where capacitor voltage V_c becomes too large and the current flow in the tube is cut off. To prevent this a *grid-leak resistor* (R_g) is shunted across the grid-leak capacitor.

It tends to discharge the capacitor at a rate that prevents a permanent charge buildup. As a result, its value must be high (2 megohms is common). Grid-leak resistors, incidentally, seem to be a high failure rate item on older radios because high value resistors tended to be unstable and were often found with extremely high, essentially open, resistance values. Now let's return to the grid-leak detector circuit.

Again let's adapt our discussion from the older Navy training manuals:

The grid-leak detector functions like a diode detector combined with a triode amplifier. It is convenient to consider the detection and amplification as two separate functions. In Fig. 8-6A the grid functions as a diode plate. The values of C_d and R_d must be chosen so that C_d charges during the positive peaks of the incoming signal and discharges during the negative peaks. The time constant of $R_d C_d$ should be long with respect to the RF cycle, and short with respect to the AF cycle.

An analysis of the waveform existing in the diode (grid circuit) is shown in Fig. 8-6B. Part 1 shows the input waveform, which is also the waveform in the input to the circuit. Because RF current i_g flows in only one direction in the grid circuit, Part 2 shows the rectified current waveform in this circuit. Part 3 shows the waveform developed across C_d. The audio waveform is produced in the same manner as the audio waveform in the diode detector circuit. However, the waveform shown in Part 3 is the output voltage. The grid-leak detector waveform produced across C_d is combined in series with the RF waveform in the tuned circuit to produce the grid-to-cathode waveform shown in Part 4.

An approximate analysis of the waveforms existing in the triode plate circuit is shown in Fig. 8-6C. Part 6 is the plate current waveform and Part 5 is the plate voltage waveform. Capacitor C1 discharges on the positive half-cycles of grid input voltage. The discharge path is clockwise through the circuit, including the tube and capacitor. The time constant of the discharge path is the product of the effective tube resistance and the capacitance of C1. This time constant is short because the effective resistance of the tube is low. The increase in plate current is supplied by the capacitor rather than the V_{bb} supply, thus preventing any further increase in current through the RF choke and plate load resistor R_L. Therefore, any further change in plate and capacitor voltage is limited.

Capacitor C1 charges up as plate voltage rises on the negative half-cycles of RF grid input voltage. The charging path is clockwise through the circuit containing the capacitor, RF choke, load resistor R_L and the V_{bb} supply. The rise in plate voltage is limited by the capacitor charging current which flows through the RF choke and through R_L. The plate current decrease is approximately equal to the capacitor charging current; thus, the total current through the RF choke and R_L remains nearly constant, and the plate and capacitor voltage rise is checked.

Positive grid swings cause sufficient grid current to produce grid-leak bias. Low plate voltage limits the plate current on no signal in the absence of grid bias. Thus, the amplitude of the input signal is limited, since with low plate voltage the cutoff bias is low, and that portion of the input signal that drives the grid voltage below cutoff is lost. The waveform of the voltage across capacitor C1 is shown by Part 6 of Fig. 8-6C. The plate voltage ripple is removed by the RF choke (RFC). Part 7 shows the output voltage waveform. This waveform is the difference between the voltage at the junction of R_L and RFC with respect to the negative terminal of V_{bb} and the voltage across coupling capacitor C_c, which for most practical purposes is a pure dc voltage.

Because the operation of the grid-leak detector depends upon a certain amount of grid-current flow, a loading effect is produced which lowers the selectivity of the input circuit. Recall that placing a resistance load in parallel with a tuned circuit widens the bandwidth. The loading which occurs in this circuit effectively places such a resistance in parallel with the input tuned circuit. However, the sensitivity of the grid-leak detector is moderately high for low-amplitude signals, and this partially offsets the disadvantages of lowered selectivity.

Regenerative detectors

Early in the development of radio, Edwin H. Armstrong produced the extremely sensitive, if a bit unstable, detector circuit called the *regenerative detector*. The circuit of a simple regenerative earphone radio is shown in Fig. 8-7A, while the equivalent circuit as used in a multitube radio is shown in Fig. 8-7B.

The regenerative detector produces a small feedback signal between the plate circuit and grid circuit. Armstrong provided this signal by means of a third inductor (L1C) in close proximity to the input coil (L1A) and tuned circuit (L1B/C1).

Called a *tickler coil*, L1C is usually either wound on the same coil form as the two other coils or is mounted inside L1B in such a manner as to allow the operator to adjust the degree of coupling between the two coils.

When the phasing of the tickler coil is correct and the feedback signal amplitude is sufficient, the feedback signal will excite the input tuned circuit L1B/C1 and cause the triode to oscillate at the resonant frequency of the tuned circuit. In the regenerative detector circuit, the grid-leak bias prevents the circuit from breaking into oscillation until the input RF signal peaks overcome the bias and allow the circuit to briefly break into oscillation.

When the oscillation heterodynes with the input RF signal, a beat note equal to the difference between the two frequencies is heard in the headphones. Hence, the regenerative detector is best used for Morse code CW transmissions. Some amateur radio operators who have a *Q multiplier* (a form of regenerative detector) on their more complex receivers use this phenomenon even today to reconstruct CW signals in the face of the large amounts of "QRM" (interference).

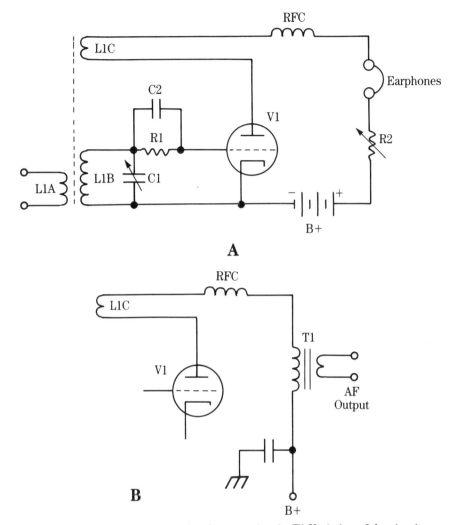

8-7   (A) Armstrong regenerative detector circuit. (B) Variation of the circuit.

The version of the regenerative detector shown in Fig. 8-7A is actually a complete radio, and this circuit was once popular especially with amateur builders. The version of Fig. 8-7B is a bit more complex and offers the ability to couple the AF signal to following stages for AF amplification.

Modern diode detectors

The diode detector of Fleming's time was not well received, and according to many extant period accounts, was inferior to a well-selected galena crystal detector. But vacuum tube development continued, and the diode tube became the envelope detector of choice for most AM demodulators in practical radio receivers.

In some cases, the tube was a single diode, while in others it was a dual diode. Two diode elements were used in the same glass envelope.

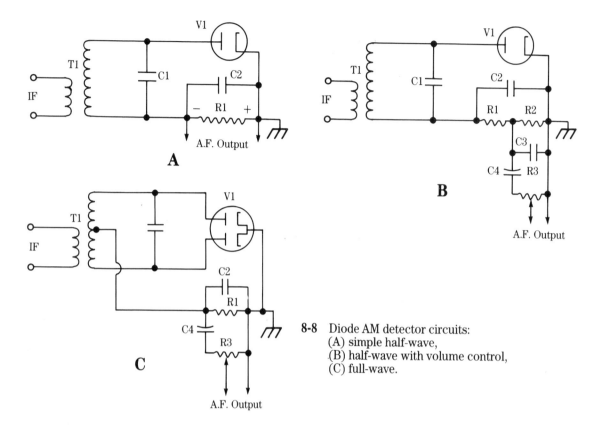

A

B

C

8-8 Diode AM detector circuits:
(A) simple half-wave,
(B) half-wave with volume control,
(C) full-wave.

The newer diode detectors tended to be used in superheterodyne receivers, although examples of TRF use of diodes was also known. In the superhet radio the plate of the diode detector is connected to the output of the last IF transformer (Fig. 8-8A), while the diode cathode is grounded.

This connection forms a half-wave rectifier for the IF signal from the IF transformer. This half-wave signal is filtered by resistors and capacitors, and then applied to the volume control to form the AF output to the following stage (which will be an audio preamplifier).

Figure 8-9 shows the waveforms to expect in the diode detector circuit. The modulated signal from the output of the IF amplifier (Fig. 8-9A) is applied to the diode rectifier to form the half-wave rectified example shown in Fig. 8-9B. The modulating envelope is still clearly seen, but the RF pulses produced by the half-wave rectification process are still present.

Capacitor C2 in Fig. 8-8A filters out the 455 KHZ IF component, producing a varying dc signal that represents the modulation envelope (Fig. 8-9C). This signal is essentially the audio signal with a dc offset from the zero baseline.

The AF signal with dc offset is still not acceptable, so some means must be provided to remove the dc offset. In Fig. 8-8B the modified half-wave diode detector uses a coupling capacitor (C4) to remove the dc offset and produce the true audio ac waveform shown in Fig. 8-9D.

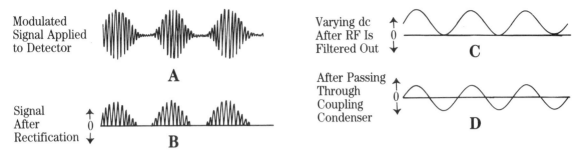

Modulated Signal Applied to Detector

A

Signal After Rectification

B

Varying dc After RF Is Filtered Out

C

After Passing Through Coupling Condenser

D

8-9 AM detector waveforms.

A full-wave rectified detector is shown in Fig. 8-8C. This circuit works in the same manner as the half-wave circuit, but produces a higher frequency output IF component of the detected waveform.

While the IF component of the half-wave detected waveform is the IF frequency, the full-wave version produces a component equal to $2\times$IF, so it is easier to filter out. Tube V1 in Fig. 8-8C might be a 6H6 or 6AL5, both of which are dual detector diodes.

Solid-state diode detectors

In more recent radios, the diode tube is replaced by solid-state pn-junction diodes. Figure 8-10 shows some solid-state diode detector circuits used in AM radios. Figure 8-10A uses a single diode in a half-wave rectified circuit. This circuit is

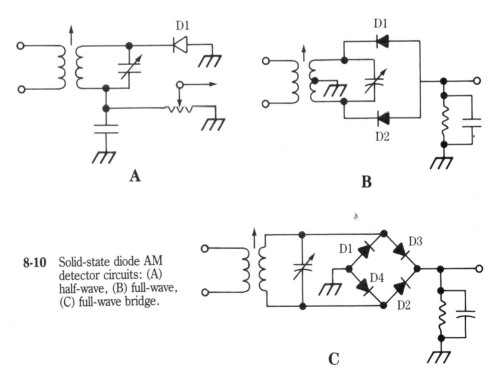

8-10 Solid-state diode AM detector circuits: (A) half-wave, (B) full-wave, (C) full-wave bridge.

analogous to the vacuum tube version of Fig. 8-8A. A full-wave rectified version is shown in Fig. 8-10B, and is analogous to Fig. 8-8C. The version shown in Fig. 8-10C is a *bridge-rectifier full-wave detector*, and has no counterpart in the previous discussion.

Automatic Volume Control Circuits

When you tune across the band using a radio receiver that lacks an *automatic volume control* (AVC) circuit, the large difference between the signal strengths of the various stations causes the radio output to alternately blast you out of the room and go quiet. Likewise, when listening to long-distance stations, ordinary propagation variations cause the signal level to fluctuate at any given reception site, so the output of the radio will vary with the vagaries of the ionosphere. The solution to this problem is to incorporate an AVC circuit into the radio design.

Note: Automatic volume control (AVC) is essentially the same as *automatic gain control* (AGC). Both work to keep the output signal level constant by varying the RF, IF, and often the mixer/converter gains in the receiver. The modern term is AGC, while on older radios it is AVC, but both are essentially the same.

Figure 8-11A shows how radio designers add a simple AVC circuit to the regular AM detector. The half-wave rectified signal (Fig. 8-11B) produces an audio ac superimposed on a dc offset (Fig. 8-11C) that is proportional to the signal strength. It is this dc signal that is used to form the dc AVC voltage (Fig. 8-11D) after heavy filtering in an RC network, consisting of C2, C3 and R2 (Fig. 8-11A).

In Fig. 8-11A, the normal static tube bias (V_k) in the IF amplifier circuit (V1) is provided by the cathode bias resistor. This voltage is, however, only part of the required bias. The other component of the bias voltage on V1 is the AVC voltage, V_{avc}.

The total grid bias is the algebraic sum of V_k and V_{avc}. Because V_{avc} varies with and is proportional to the signal level at the output of the IF amplifier, the bias applied to V1 also varies. Therefore, the gain of the tube used for the IF amplifier varies inversely with the signal strength, smoothing out the fluctuations caused by different signal strengths.

In the 1930s and later, the AM detector, AVC rectifier (often the same diode), and audio frequency preamplifier were combined into one tube. The 6SQ7, 12SQ7, 6AV6, and 12AV6 tubes were dual diode with triode section, and were especially designed for this application.

Figure 8-12 shows a typical circuit in which the two diode plates are connected together to form a single diode detector/AVC rectifier. In other cases, only one diode is used, and the other diode plate is shorted either to ground or to the cathode terminal, depending upon whether or not the triode uses cathode bias. A few circuits even use the second diode plate to form a separate AVC rectifier using a signal sample from elsewhere in the circuit. The triode section is used as a voltage amplifier to provide preamplification to the AF signal before it is applied to the power output stage.

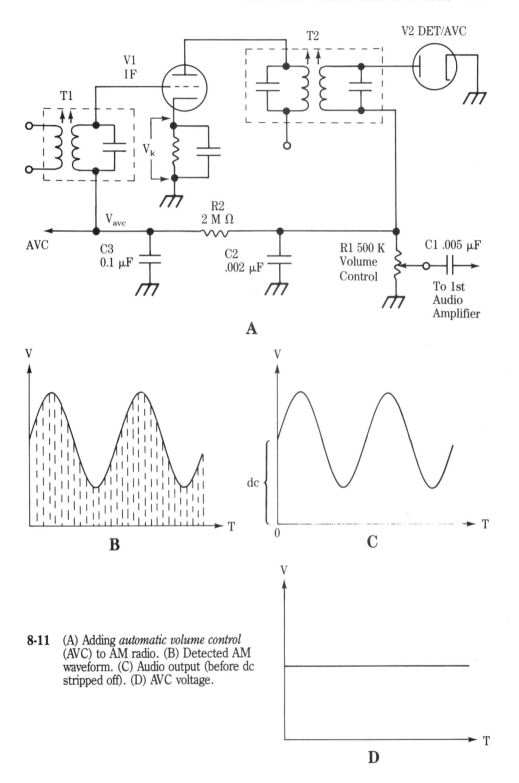

8-11 (A) Adding *automatic volume control* (AVC) to AM radio. (B) Detected AM waveform. (C) Audio output (before dc stripped off). (D) AVC voltage.

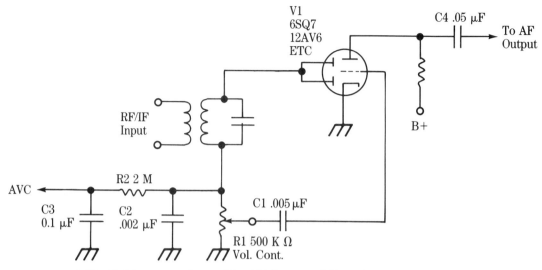

8-12   Multipurpose tube used for AM detector, AVC, and the first audio stage.

Tweet Filters

AM band channel assignments in the United States (and most of the world) are spaced 10 KHz apart. American AM band channels typically end in zero: 790 KHz, 800 KHz, 810 KHz, etc. The Federal Communications Commission will not assign two stations to adjacent channels in the same region because they will interfere with each other. However, adjacent channels, 10 KHz apart, can be heard in the same area even though one or both stations might be many miles away.

Also, foreign stations in Europe and Latin America are often found on what we in the United States call *split channels*. These channels are spaced 10 KHz apart, but are on the odd frequencies ending in 5, e.g., 885 KHz and 995 KHz, instead of the 880 KHz and 990 KHz used in the United States.

Consider the scenario shown in Fig. 8-13A. In this case, a U.S. station on 880 KHz is being received simultaneously with an 885 KHz station from a "border blaster" station. (If you're old enough to remember what a "border blaster" is, congratulations.) The two signals are both within the selectivity passband of the receiver, so they will heterodyne together to produce a 5 KHz beatnote in the output. Because this 5 KHz signal is within the audio passband of the receiver, it will be heard by listeners—who'll curse the interfering station.

Similarly, unless the audio passband is limited, adjacent channel stations in the U.S., spaced 10 KHz apart, will create a 10 KHz beatnote in the output. This phenomenon is the cause of the whistles and shrill sounds normally heard in the outputs of most receivers. These whistles are called *tweet signals* in the jargon of radio.

The solution to the tweet signal, especially the 10 KHz variety, is to incorporate an audio low-pass filter into the output of the detector circuit. This filter is shown as C1/C2/R1 in Fig. 8-13. This audio low-pass filter is usually called a *tweet filter* in radio manuals.

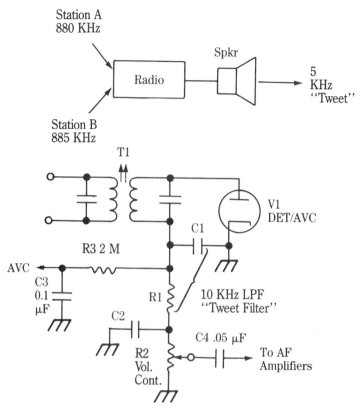

8-13 (A) Source of squeals on the AM band—two signals close together. (B) "Tweet filter" to remove the squeal.

9
Radio Repair Bench Test Equipment

SOME PARTICULARLY SKILLED TECHNICIANS SEEM TO BE ABLE TO DIVINE THE PROBLEM IN a radio receiver. While it seems like magic to the noninitate, it is really only a reflection of the large amount of experience at work; and the technician really only remembered from a past service job. But even the most skilled, seemingly magical technician will agree that a good selection of test equipment is the key to wringing out problems and repairing faults.

Figure 9-1 shows my own workbench, although since the photo was taken I've added two additional signal generators, a spectrum analyzer, and a couple of other small devices. Note that the test equipment is a mix of old and new. The large black signal generator in the center is a Measurements Model 80 that was made in 1949, which I refurbished. It produces up to 100,000 µV of RF, modulated or unmodulated, at frequencies from 2 MHz to 400 MHz.

I traded a local ham a "parts-only" sideband rig for the signal generator. He rebuilt the SSB rig and I rebuilt the Model 80 (and we both knew the condition of the other's offering, lest you wonder). At hamfests I've seen these same generators in both military and civilian versions for as little as $50.

The nature of the test equipment you select depends on the job you want to do. If all you want to do is service *art deco* AM table model radios, then a simple 1947 *Precision E-200C* and a VTVM are all you need for most jobs.

In my own case, I need a larger collection because of my writing interests. My interest in classic radios, amateur radio, and electronic hobbyist interests all need a different (if overlapping) selection of equipment. Figure 9-1 is the result.

For classic radio service you will need several different types of instruments: a multimeter, signal generator, oscilloscope, power supply, and certain others, which we will deal with shortly. Let's take a look at these instruments, each in its turn.

9-1   Typical radio repair bench.

Multimeters

A *multimeter* is an instrument that combines into one package a multirange voltmeter, milliammeter (or ammeter), ohmmeter, and possibly some other functions. The single instrument will measure all the basic electrical parameters the technician needs to understand in troubleshooting a receiver.

The VOM

The earliest multimeters were instruments called *volt-ohm-milliammeters* (VOMs). These instruments were passive devices; the only power used in them was a battery for the ohmmeter function. Because they are passive instruments they are still preferred for troubleshooting medium- to high-power transmitters. Active instruments contain electronic components that can be affected by electromagnetic fields produced in transmitters.

The sensitivity of the VOM was a function of the full-scale deflection of the meter movement used in the instrument. Sensitivity was rated in terms of ohms per volt, full-scale. Three common standard sensitivities were found: 1000 ohms/volt, 20,000 ohms/volt and 100,000 ohms/volt. These represented full-scale meter sensitivities of 1 mA, 10 µA, and 100 µA, respectively.

Voltage was read on the VOM by virtue of a "multiplier" resistor in series with the current meter, which forms the basic instrument movement. The front panel switch selected which multiplier resistor, hence which voltage scale was used.

The VTVM

The earliest form of active multimeter was the *vacuum tube voltmeter* (VTVM). This instrument uses a differential balanced-amplifier circuit based on dual triode tubes (e.g., 12AX7) to provide amplification for the instrument. As a result, the VTVM was more sensitive than the VOM. It typically had an input impedance of 1 megohm, with an additional 10 megohms in the probe, for a total input impedance of 11 megohms. Compare this feature with the VOM, which had a different impedance for each voltage scale.

The DMM

The next improvement in meters was the solid-state meter. These instruments used a field-effect transistor in the front-end, and other transistors in the rest of the circuit. As a result, these instruments were called *field-effect transistor voltmeters* (FETVMS). The FETVM has a very high input impedance, typically 10 megohms or more.

Finally, the most modern form of instrument is various *digital multimeters* (DMMs) in both handheld and bench (Fig. 9-2). These instruments are the meter of choice today if you are going to purchase a new model. In fact, it might prove a little difficult to find a nondigital model anyplace except Radio Shack.

The DMM is preferred today because it provides the same high-input impedance as the FETVM. But, because the display is numeric instead of analog, the readout is less ambiguous. There is one bit of ambiguity, however, called *last-digit bobble*. The digital meter only allows certain discrete states, so if a minimum value is between two of the allowed states, then the reading might switch back and forth between the two, especially if noise is present. For example, a voltage might be 1.56456 volts, but the instrument is only capable of reading to three decimal places. Thus, the actual reading might bobble back and forth between 1.564 and 1.565 volts.

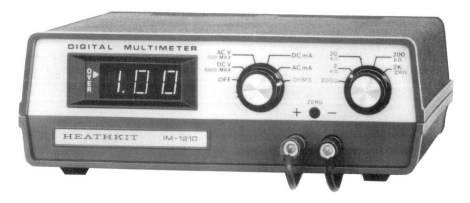

9-2 Digital multimeter.

Last-digit bobble is not really a problem, however, unless you bought an instrument that is not good enough. The number of digits on the meter should be one more than that required for the actual service work. The way the digits are specified in DMMs is in terms of the number places that the meter can read. For example, the instrument in Fig. 9-2 is considered a 2 1/2-digit model. The "1/2" digit refers to the fact that the most significant digit (on the left side) can only be 1 or 0 (if it is a zero it is blanked off). Thus, the full scale of this meter is always 199. Where the decimal point is placed is a function of the range setting.

For most vacuum tube radio service, the DMM selected can be a 2 1/2-digit model, but if you anticipate solid-state radio servicing, then opt instead for a 3 1/2-digit model. These instruments will read to 1999 on each scale. Oddly, the cost difference between the two is small.

An interesting aspect of the DMM is that they use low voltage for the ohmmeter, so the instrument can be used in the circuit on solid-state equipment. The ohmmeter on other forms of instrument will forward-bias the pn-junctions of solid-state devices, so it cannot be used for accurate measurements.

However, this same feature of the DMM also means that the normal ohmmeter cannot be used to make quick tests of pn-junction devices, although some DMMs have a special switch setting that will allow this operation. In some it is a "high-power" setting, while in others it is designated by the normal arrow diode symbol.

Another feature of the modern DMM is an aural continuity tester. It will sound a beep anytime the resistance between the probes is low, indicating a short circuit exists. This feature is especially useful for testing multiconductor cables for continuity when the actual resistance is not a factor.

Older Instruments

As is true with other equipment, the multimeter shows up on the used market (especially hamfests) quite often. It is easy to obtain workable, if older, instruments cheaply. This is especially true now that everyone seems to need to dump old analog instruments in favor of the new digital varieties. However, there are some pitfalls.

First, make sure that the instrument isn't so old that it uses a 22 1/2-volt (or higher) battery in the ohmmeter section. Those instruments were made prior to about 1956, and will blow any transistor that you measure with them. Look out especially for the oversized RCA and Hickcok instruments.

Second, make sure that there is nothing wrong with the instrument that cannot easily be repaired. VTVMs tend to be in good shape, or at least repairable condition. Unfortunately, a lot of VOMs are in terrible shape.

Part of the problem comes from the fact that the meter will measure current easily enough, but if the operator left the meter on the current setting and then measured a voltage—pooff! The instrument burned up. When examining an instrument, try to make a measurement on each scale. Alternatively, remove the case and look for charred resistors and burnt switch contacts.

In any event, when you obtain a used meter, be sure to take it out of the case, clean the switch contacts, and generally spruce it up. The usual contact and tuner cleaners will work wonders on the used multimeter as well. A simple cleaning of the switch(s) might remove intermittant or erratic operation.

Signal Generators

The purpose of a signal generator is to produce a signal that can be used to troubleshoot, align, or simply improve the performance of a piece of electronic equipment. Fortunately for radio service, only a few basic forms of signal generator are needed. Signal generators for this market include audio generators, function generators, and RF signal generators.

Audio generators

Audio generators produce sinewaves on frequencies within the range of human hearing (20 Hz to 20 KHz). Some models produce frequencies over a greater range, while others produce square waves as well as sinewaves. The standard audio signal generator has a variable amplitude or level control, and a fixed output impedance of 600 ohms. Audio generators may or may not have an output level meter.

Figure 9-3 shows a simple *Heathkit* audio signal generator that can be built from a kit (there is also an RF generator in the same series). This small instrument produces sine- and square-wave outputs at frequencies from 10 Hz to 100 Hz. The sine- and square-wave outputs are independent of each other. This instrument meets the basic needs of radio service work, but will not suffice for serious high fidelity measurements. It will, however, work nicely for hi-fi troubleshooting.

A more complex audio signal generator is the *Heath* IG-18 shown in Fig. 9-4. This instrument produces sine- and square-wave outputs that can be individually calibrated for amplitude. The meter measures the output level of the sinewave signal.

Frequency is set on this instrument by using a series of switches. The first switch is a range multiplier, while each successive switch is an order of magnitude

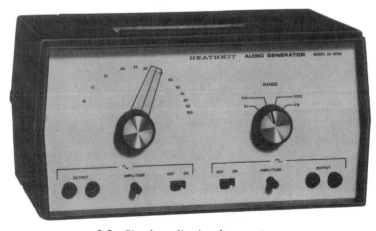

9-3   Simple audio signal generator.

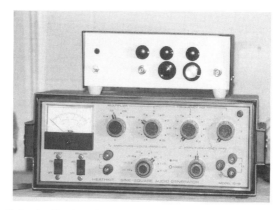

9-4   Audio signal generator with measurable output.

less (but arranged in decade groupings). The settings shown are ×10, 30, 0, 1, so the frequency set is 10 × (30 + 0.1) or 301 Hz. The box on top of the IG-18 signal generator is a homebrew-project that will produce sawtooth and triangular waves from the IG-18 outputs. Since this photo was taken, the circuitry in the small box was added to the IG-18 inside the case.

Function generators

A function generator (Fig. 9-5) is much like an audio generator, but puts out a triangular waveform in addition to the sine- and square-wave signals. Some function generators also produce pulses, sawtooths and other waveforms as well. Typical

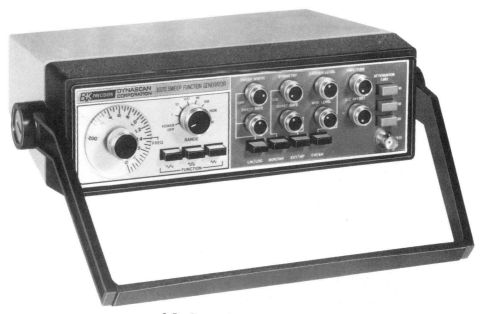

9-5   Sweep function generator.

function generators operate from less than 1 Hz to more than 100 KHz. Many models exist that have a maximum output frequency of 500 KHz, 1 MHz, 2 MHz, 5 MHz, and in one case 11 MHz. Like the audio generator, the standard output impedance is 600 ohms.

However, some instruments also offer 50 ohm outputs (standard for RF circuits), and a TTL compatible output. The latter is a digital output that is compatible with the ubiquitous *transistor transistor logic* (TTL or T^2L) family of digital logic devices (these are the ones with either 74xx or 54xx type numbers).

The *B&K Precision* Model 3020 shown in Fig. 9-5 is a full function generator, but also adds sweep. The sweep generator allows the output frequency to sweep back and forth across a range around the set center frequency. As a result, one can do frequency response evaluations using the sweep-function generator and an oscilloscope.

Figure 9-6A shows the swept frequency output of a sweep-function generator, while Fig. 9-6B shows the same signal at the output of a low-pass filter circuit. The frequency response of the low-pass filter can easily be seen.

RF Signal Generators

RF signal generators put out signals generally in the range above 20 KHz, and typically have an output impedance of 50 ohms. This value of impedance is common for RF circuits except in the TV industry, where 75 ohms is standard. RF signal generators come in various types, but can easily be grouped into two general categories: service grade and laboratory grade. For service work, either type is usable. Figure 9-7 shows a *B&K Precision* Model 2050 service grade instrument that is moderate in cost, but meets the needs of most users who need to service radio receivers.

Although those who service high-grade communications receivers and transceivers will not find the instrument useful for alignment purposes, all other servicers will find it more than ample. It is the descendent of a long line of instruments of the same series back to the 1940s (see chapter 25).

A   **B**

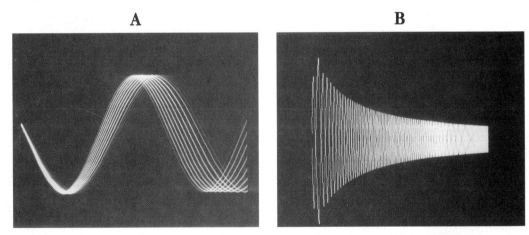

9-6   Sweep generator output: (A) direct measurement, (B) into low-pass filter.

9-7 Service-grade RF signal generator.

My own bench has several different signal generator instruments: the Heath IG-18, a sweep-function generator, a Model 80 (2–400 MHz) RF signal generator, and an elderly Precision Model E-200C that I refurbished after buying it at a hamfest in nonworking condition.

Oscilloscopes

Probably no other instrument is as useful as the oscilloscope in troubleshooting radio receivers. You can look at signals and waveforms instead of just averaged dc voltages as measured on the DMM. Figure 9-8 shows a simple service grade instrument that can be used for most service jobs on AM, FM, and shortwave receivers, as well as on most CB and Amateur radio transmitters.

The oscilloscope (scope) was originally called the *cathode ray oscillograph* (CRO). It displays the input signal on a viewing screen that is provided by a cathode ray tube (CRT).

The light on the viewing screen is produced by the CRT deflecting an electron beam vertically (Y-direction) and horizontally (X-direction). When two signals are viewed with respect to each other on an X-Y oscilloscope, the result is a *Lissajous figure*, as shown in Fig. 9-9A. The pattern shown in Fig. 9-9A demonstrates a 2:1 frequency ratio between the X and Y input signals. If the X-direction is driven by a sawtooth waveform, then the display of the Y-input will be an amplitude vs. time pattern, as shown in Fig. 9-9B.

The quality of oscilloscopes is often measured in terms of the *vertical bandwidth* of the instrument. This specification refers to the -3 dB frequency of the vertical amplifier (or amplifiers if it is a dual-beam scope). The higher the bandwidth,

9-8 Service-grade oscilloscope.

the higher the frequency that can be displayed and the sharper the rise-time pulse that will be faithfully reproduced.

For ordinary service work, a 5 MHz scope will suffice for most applications. However, there will be times when you will need a high-frequency scope, so buy as much bandwidth as you can afford. Although once in the price stratosphere, even 50 MHz models can be bought relatively cheaply today—brand new.

A   **B**

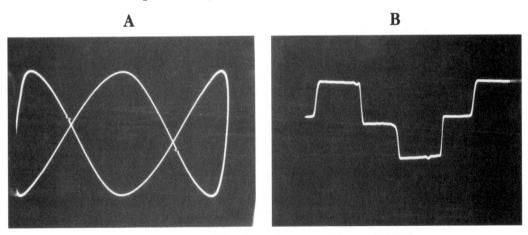

9-9 Oscilloscope patterns: (A) Lissajous figures, (B) V-T.

9-10   dc power supply.

DC Power Supplies

Although often overlooked as "test equipment," the simple dc power supply (Fig. 9-10) is definitely part of the test bench. In addition to powering units being repaired that either lack a dc supply or have a defective dc supply, the bench power supply can be used for a variety of biasing and other troubleshooting functions.

The model shown in Fig. 9-10 is a 0- to 20-volt dc supply that will produce up to 500 mA. This supply is useful for all solid-state radios except car radios and hi-fi sets (both of which require more current). Notice in Fig. 9-10 that there are two outputs ("−" and "+") and that the ground connection is separate.

While one might think that the ground and the negative ought to be connected together, well thought out designs keep them separate. The reasons include both safety and versatility. When the power outputs are floating, there is less chance of an accidental short to ground. In addition, two or more supplies can be connected in series for higher-voltage operation if the outputs are floated.

Other dc supplies available include dual-polarity models, high-current models, high-voltage models and so forth. You will be able to find almost any form of dc supply that you need.

Other Instruments

Signal tracer   The signal tracer is a high-gain audio amplifier that can be used to examine the signal at various points along the chain of stages in a radio or audio amplifier. A *demodulator probe* will permit the signal tracer to "hear" RF and IF signals.

The signal tracer is a backup replacement for the oscilloscope if it is absolutely impossible to obtain a scope. However, the signal tracer is such a poor backup to the scope it is considered insufficient. However, there are advantages to these instruments, and their cost is so low that they are a worthwhile addition to the bench.

Distortion analyzer   Figure 9-11 shows two forms of distortion analyzer that can be used for troubleshooting high fidelity equipment, especially the classic

A

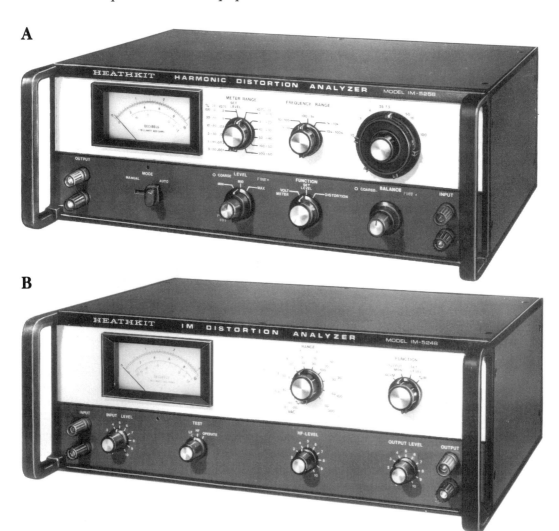

B

9-11 Total Harmonic Distortion (THD) analyzer.

variety where the levels of distortion did not approach the low levels possible today. Figure 9-11A shows a total harmonic distortion analyzer, while Fig. 9-11B shows an intermodulation distortion analyzer.

10
Basic Radio Troubleshooting

TROUBLESHOOTING IS THE ART OF EXAMINING A RADIO, OR OTHER ELECTRONIC DEVICE, AND discovering why it is not performing as it should. Once the fault is found, the radio can be repaired by replacing a defective component or fixing whatever is broken. In this chapter we will look at the basics of radio troubleshooting. This information will start you on the way to being able to do the job yourself.

Before attempting to troubleshoot a radio we need to be able to describe what kind of trouble the radio is having. It is also necessary to examine the radio to see if something is missing, broken, cracked, or otherwise out of order. If the radio works, then check the entire tuning range and note whether or not stations appear at the right spot on the dial.

If you are repairing the radio for someone else, either commercially or as a favor, then don't count on the owner knowing enough to give you an accurate description. For example, the owner might say that the radio has a "hum," so you might expect a 60 Hz or 120 Hz tone to come from the speaker. The actual noise might be a 10 KHz whistle, a putt-putt-putt motorboating type sound, or a 1000 Hz oscillation.

If you know the radio has worked recently, and then simply failed, then you know that the radio is at least complete. Examine the ac line cord (if one is used) to make sure that it is safe. Plug the set into the ac outlet. If you own an isolation transformer (and you should own one if you plan to repair radios), then make sure the radio is plugged into the isolation transformer.

Turn the set on, and set the volume control to about half-scale. On vacuum tube sets the radio might take a minute or two to come on, so don't be alarmed if the radio takes awhile to warm up—these are not transistor sets! Tune the radio from station to station and note what happens, or what doesn't happen.

Adjust the volume control back and forth through its entire range several times and note whether or not scratching is heard. This action will tell you the condition of the volume control, and whether or not the audio section is working. The electrical noise created when the control is operated is a valid signal as far as the audio amplifiers are concerned, so these stages will pass the noise on to the output if they are working properly.

Let's consider several cases of generic trouble, and go through the methods for troubleshooting those cases. We will examine the following cases: hum in the output, dead radio, weak signals, motorboating, noise, oscillations, and distortion.

Hum in the Radio Output (Radio Works Otherwise)

Hum is a low-frequency oscillation or noise in the output of the receiver. There are several different types of hum seen in radio receivers: power-supply ripple hum, 60 Hz nontunable hum, and modulation or tunable hum.

Power-supply ripple hum

Chapter 5 covered the dc power supplies used in radio receivers. One form of power supply produces high-voltage dc from the ac input from the power mains by means of a rectifier. But the output of the rectifier is not perfect dc; rather, it is pulsating dc.

A ripple filter, consisting of capacitors and either a resistor or inductor (choke), smoothes the pulsating dc into a more pure form of dc. If there is a defect in this filter, then the B+ will be modulated by the 60 Hz (half-wave rectifiers), or 120 Hz (full-wave rectifiers).

Power-supply ripple can be identified immediately in full-wave rectified cases because the hum will have a frequency of 120 Hz. If the radio uses a half-wave rectifier, however, the hum will be at 60 Hz. Some technicians can distinguish power-supply hum because it is a little more raucous sounding than other forms of 60 Hz hum. The oscilloscope will also tell the tale.

Power-supply hum is most often heard when the volume control is turned all the way down. However, it might be possible at times for it to be heard only when the radio is turned up. This condition is rare, however, and is often accompanied by whistles and oscillations. Let's consider a practical example.

Figure 10-1A shows the half-wave rectified dc power supply for an "All-American-Five" table model radio from the early- to mid-1950s. The ripple filter consists of resistor R1 and capacitors C1A and C1B. The filter capacitor is a dual tubular electrolytic capacitor (Fig. 10-1B).

The quickest way to determine whether or not the problem is a filter capacitor is to bridge a known good electrolytic capacitor across the suspected bad section (Fig. 10-2). First turn off the power and disconnect the ac line cord. Wait a few seconds for the capacitors to discharge, and then complete the discharge process by connecting a jumper cord across the capacitor (use one that is insulated except at the tips in order to prevent an electrical shock). Solder-tack the known good capacitor across the suspected bad section, making sure that its leads are not touching anything else, and then plug the radio back into the ac power. Turn the radio on and wait for it to warm up.

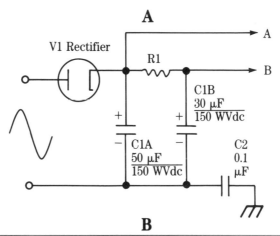

A

B

10-1 (A) Simple half-wave rectified dc power supply. (B) "All-American-Five" radio chassis.

If the capacitor is bad, then the radio will work normally (assuming no other faults elsewhere in the radio). To repair the radio, turn the power off and unplug the set. Again discharge the capacitors and unsolder the temporary connections.

Replacing the bad capacitor is the only solution, even though some electrolytic capacitors seem to "heal" themselves on occasion. Dual tubular electrolytics are hard to find today, but not impossible. If you prefer, it is also reasonable to use two

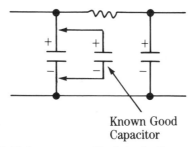

Known Good
Capacitor

10-2 Bridging suspected bad ripple-filter capacitor.

single-section electrolytic capacitors as the replacement. In one radio that I repaired recently the "50 µF/150 WVdc + 30 µF/150 WVdc" dual capacitor was replaced with 47 µF/165 WVdc tubular and a 35 µF/165 WVdc units, respectively.

Some people attempt to repair a multisection electrolytic capacitor by permanently bridging a single electrolytic capacitor across the defective section. This is always a bad practice, so don't do it!

An oscilloscope can also be used for locating the defective section. Use the scope in the ac-coupled input mode, and adjust the vertical attenuator to show a reasonable trace on the screen. Figure 10-3 shows the before and after waveforms at point A, across capacitor C1A. Figure 10-4 shows the waveforms at point B, across capacitor C1B.

The radio in question had not been turned on in many years, so both sections of the dual electrolytic capacitor were open. Figure 10-3A shows the waveform across input filter capacitor C1A when C1A was defective. Note the extremely high ripple. When the new replacement filter was installed, the ripple dropped to the level shown in Fig. 10-3B.

Likewise, in Fig. 10-4A we see the ripple waveform across C1B when the filter capacitor was open, while in Fig. 10-4B shows the waveform when both filter capacitors are working properly.

60 Hz nontunable hum

The nontunable form of 60 Hz hum, assuming that it is not caused by a defective filter capacitor in the dc power supply, is most often caused by cathode-to-heater or cathode-to-grid (directly heated cathode tubes) short circuits in the vacuum tubes. Perhaps the most common form is the hum heard when the volume control is turned down.

In this case, the most likely suspect is the audio output (AF power-amplifier) stage. Check the circuit by temporarily replacing the tube. In other cases, the cathode short will be in another tube, with slightly different effects on the output. A vacuum tube tester with the ability to detect short circuits can be used to locate the

A   **B**

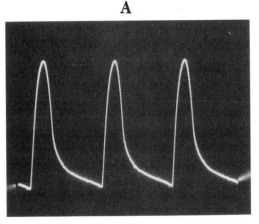 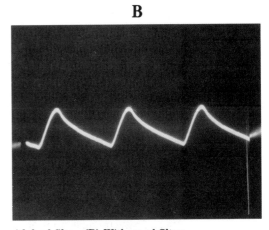

10-3   (A) Ripple at output of rectifier with bad filter. (B) With good filter.

A **B**

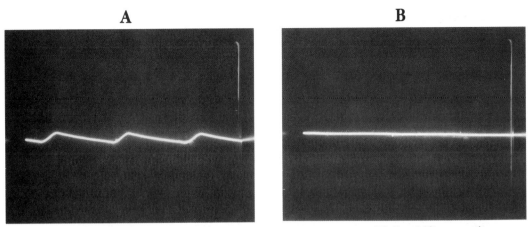

10-4 (A) Ripple at output of RC ripple filter with bad capacitor. (B) Good filter capacitor.

bad tube, even though the emission and transconductance of the tube may be unaffected.

Tunable hum

This form of hum, also called *modulation hum*, is heard only when the radio is tuned to a radio station, and might disappear when the announcer quits speaking (although that effect is only seen occasionally).

There are two main causes of modulation or tunable hum. First, the ac line RF filter capacitor (C1 or C2 in Fig. 10-5) might be open. Replace this capacitor with an identical unit of the same value.

Note: the ground end of these capacitors is marked with a stripe or dot. The term *ground end* does not refer to a polarity situation, but rather to the lead that connects to the outer plate of the capacitor. Noise and RF interference tend to be less when the ground end is connected to ground.

The second possible cause of tunable hum is a cathode-heater short in the RF amplifier or converter tube (or, on older sets, the local oscillators and/or the mixer).

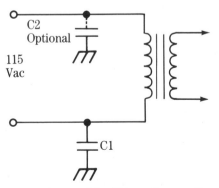

10-5 Antihum capacitors on primary side of transformer. Capacitors selected for this application should be rated at 1,600 volts, and be specified for continuous ac operation.

Dead Radio

Probably no problem is easier to troubleshoot than the dead radio. Although this problem tends to scare customers because they believe that the problem will result in a large repair bill, the problem is much easier to solve than certain other problems and is often cheap to repair. Finding the fault is a relatively simple, logical procedure.

Check all the tubes   Most of these problems seem to result from defective tubes. Use either an emission or transconductance tube-tester for this chore. Alternatively, check the tubes by replacing them one by one with known good tubes. If the radio is operated from a power transformer, then check to see that *all* of the tubes are lighted.

A dead tube filament in this type of radio is a giveaway. Keep in mind, however, that some tube filaments will not show up even when normal unless the room lighting is either blocked or subdued. On ac/dc radios (such as the All American Five) the tube filaments are wired in series, so an open filament will turn off all the other filaments as well.

Check out the ac and dc power supplies   Once it is known that the tubes are good, then it is time to check out the power supply. Look for and check the following:

- Open line fuse
- Open ac line switch
- Open line cord
- Open power-supply transformer primary or secondary
- Open filter choke, filter resistor (R1 in Fig. 10-1A), or speaker field coil
- Physical short to ground
- Shorted filter capacitors (look for defective or sparking rectifier tube)
- Open voltage divider resistor
- Open line cord or ballast resistor

Most of these defects are self-explanatory and easily checked with either a voltmeter or ohmmeter. *Warning:* Before using an ohmmeter in a circuit with filter capacitors, make sure that the capacitors are discharged multiple times. Otherwise, damage can occur to the meter.

The last item in the list refers to ballast resistors used in some series filament circuits. Figure 10-6A shows how such a resistor is used. It is connected in series with the filaments in order to drop some of the voltage. Suppose all of the tubes in the circuit use 12-volt filaments. Five tubes will drop 5×12 volts, or 60 volts. It is necessary to drop an additional 55 to 60 volts in order to prevent the tubes from burning out.

There are two common forms of ballast resistor. In some cases it will be an ordinary power resistor, so it is easily replaced. In other cases, the resistor is a third wire in the ac line cord (Fig. 10-6B). In normal operation these line cords get quite warm, and that heat is a sign that they are working normally.

When you see an old radio that has a two-prong plug, but a three-wire line cord, then assume that the third wire is a resistance element, *not* a ground wire, as is usually true when a three-prong plug is married to a three-wire line cord.

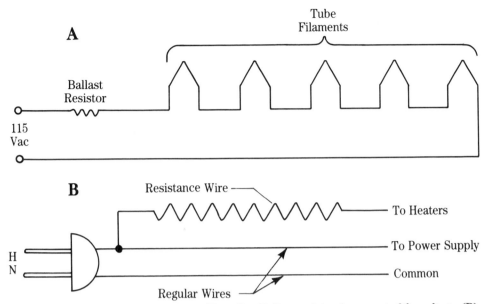

10-6 (A) Series wired filaments in ac/dc radios. Ballast resistor drops part of the voltage. (B) Line cord ballast resistor.

If a resistance line-cord resistor element is open, then the filaments in the radio's tubes will not light up. In some cases it might be possible to repair the defective line cord, but in most cases the old line cord will be either frayed or dry-rotted, so it is too dangerous to reuse.

If the line cord must be replaced, then attempt to find a replacement resistance line cord. Otherwise, use a power resistor as in Fig. 10-6A. The value of the power resistor can be found by tearing open the line cord and measuring the element. If that is possible, then the approximate value can be found from the following procedure:

1. Look in the tube manual to find the filament current of the tubes used in the radio. Many tubes use 450 mA (0.45 A) filaments, so we'll use that value in this example.
2. Look at the type numbers of the tubes. The first digits (before the letters), e.g., the "12" in "12BA6", is the filament voltage. Add up all of these voltages. In our example assume a total of 60 volts.
3. Subtract the calculated sum of filament voltages from 115 volts: $115-60 = 55$ volts.
4. Use Ohm's law, $R = V/I$, to calculate the resistance: $R = V/I = (55 \text{ volts})/(0.45 \text{ A}) = 122$ ohms. (Use 120 ohms or 125 ohms.)
5. Calculate the power dissipation in order to decide on the resistor power rating: $P = EI = (55 \text{ volts}) (0.45 \text{ A}) = 24.75$ watts. This value is less than 25 watts, but using a 25-watt resistor would be disastrous to future reliability. Select a 50-watt or larger resistor, or a combination of two or more lesser resistors to arrive at the correct rating. For example, a pair of 250 ohm, 20 watt resistors in parallel will yield a total rating of 125 ohms at 40 watts.

Check the loudspeaker and related circuit   A fault in the loudspeaker can cause the radio to be dead. Check the speaker for an open voice coil. This test can be made by disconnecting the leads from the output transformer to the speaker. If the speaker is good, then check the output transformer secondary winding with the ohmmeter before resoldering the wire to the speaker voice coil terminals.

An open loudspeaker can often be detected by ear. Listen closely (without a lot of room noise) while tuning the radio across the band. The laminations of the output transformer will vibrate in step with the audio, giving rise to audible "transformer talk." If you hear it, then either the speaker is bad or the secondary winding of the audio output transformer is open.

Finding the Defective Stage

At this point it is necessary to decide which of two troubleshooting approaches to take: signal tracing (Fig. 10-7) or signal injection (Fig. 10-8). Although there are places where one will work better than the other, there are also many other cases when the selection is a matter of preference, or simply what test equipment is available at the time.

Signal tracing   Signal tracing (Fig. 10-7) uses either a strong local on-the-air station or a signal generator as the test signal source. In Fig. 10-7 the signal generator is shown coupled to the antenna by a gimmick, although direct connection is also permissable.

When the signal source is the antenna, RF amplifier, or any other point prior to the converter stage, the radio will have to be tuned to the signal source frequency (which might take some doing if the radio is dead). However, the noise heard between stations is also a reasonable test signal "in a pinch."

Attach the probe of the oscilloscope or signal tracer to the outputs and inputs of each stage in sequence until the defective stage is found.

For example, if signal is heard at the grid of the IF amplifier tube, but not at the

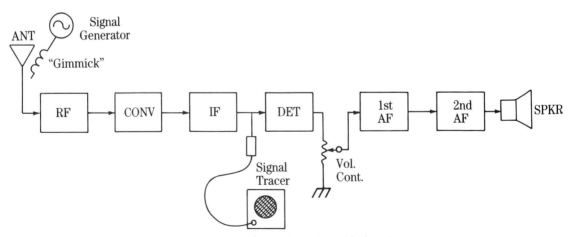

10-7   Signal tracing method of troubleshooting.

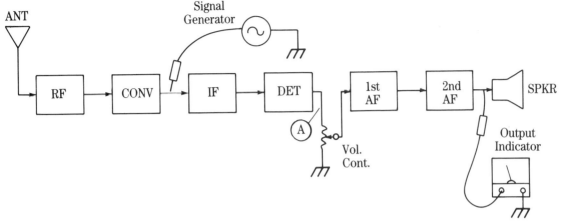

10-8 Signal injection method of troubleshooting.

plate, then the tube is not amplifying. Check the tube, the IF transformer, and the voltages applied to the tube to discover the problem. In signal tracing it is common practice to start at the front-end of the radio and work towards the speaker.

Signal injection   Signal injection (Fig. 10-8) uses a signal generator to locate the bad stage. An output indictor (oscilloscope or ac voltmeter) is connected across the loudspeaker, at least "officially." Unofficially, it is quite permissible to use the loudspeaker and your ear as the output indicator, regardless of what certain old textbooks proclaim. The signal generator will have to be an AF/RF type, offering both types of output in order to handle all stages.

Inject a signal from the signal generator into the input of the last stage in the chain, and then work backwards towards the front-end in order to find where the signal stops passing through the system.

In both signal tracing and signal injection it is a good idea to start at the volume control most of the time. This procedure divides the radio about in half, giving you an equal chance of finding the problem in either direction.

Turn the volume control up to maximum, and then inject an audio signal to the top of the volume control (point "A" in Fig. 10-8). If signal is heard in the output, then the problem is prior to the volume control. If no signal is heard, then the problem follows the volume control.

Weak Signal Reception

If the radio output signal level is weak, there are two different types of symptom. Your approach to troubleshooting depends upon which is found. First, there is the case where few signals are heard. Some local stations might be loud, but not as loud as they are normally heard.

This type of symptom points to a problem between the antenna terminals and the input of the IF amplifier. The root problem is *sensitivity*, and could result from any number of defects in the RF amplifier, converter, or input side of the IF amplifier.

In the other case, the output level is low, but many stations are heard. This symptom points to a defect from the output of the RF amplifier to the loudspeaker. The problem here is not sensitivity, but rather the gain of stages beyond the IF, or the integrity of the signal path in that same region.

Common causes of these problems include open coupling capacitors in the audio amplifier stages, open cathode bypass capacitors in the audio amplifiers, and open capacitor in the tweet filter or detector, or a defective output IF transformer.

Keep in mind: Alignment is never a troubleshooting technique. Many radios are seriously misaligned by well-intentioned repairers who are trying to find the source of weak reception. Other methods will find the bad component. Although there are radios in which alignment is the cause of weak reception, those cases almost invariably result from someone having misaligned the set in the past. Rarely does radio alignment deteriorate to the point of weak reception, especially if the failure was sudden, rather than gradual.

Standard sensitivities and stage gains can be used to troubleshoot a weak radio. Sensitivity is measured by noting the input signal level required to produce a standard output level. That standard output level (for 4 ohm speakers) is 400 millivolts when the signal generator is modulated with either 400 Hz or 1000 Hz. Rule-of-thumb sensitivity values are:

Antenna terminal	5 to 13 μV (600 KHz)
Converter grid	50 μV (600 KHz)
IF amplifier grid	3000 to 4000 μV (455 KHz)
1st AF grid	0.030 to 0.040 volts (400 Hz)
2nd AF grid	1.5 to 1.8 volts (400 Hz)

The average stage voltage gains in a typical radio range from a low to a high (typical shown in parentheses):

RF amplifier	20 to 40 (25)
Converter	60 to 80 (75)
IF amplifier	80 to 130 (100)
1st audio	40 to 60 (50)
2nd audio	5 to 20 (10)

The following list gives some of the more common trouble areas to inspect when confronted with a radio that shows a weak output signal:

- Bad (but not totally dead) vacuum tube
- Shorted power transformer (low B+ or filament voltage)
- Open cathode bypass capacitor in any stage
- Open AVC/AGC bypass capacitor
- Open plate bypass capacitor in RF/IF/converter stages
- Open antenna or IF coil
- Jammed or "bottomed out" speaker voice coil

Motorboating

The term *motorboating* refers to a low-frequency oscillation that makes a putt-putt-putt sound, resembling a motorboat engine at idle. There are four main causes of this problem, although not all are seen on all receivers. On vacuum tube radios, the principal causes are open power supply filter capacitors, and an open grid circuit.

A third cause, seen in transistor portable radios made in the 1950s and 1960s, is low battery energy. When the battery is almost depleted, its internal resistance rises and decoupling becomes ineffective. A fourth cause, seen in transistor car radios made up to about 1970, is an open AGC filter/decoupling capacitor.

Motorboating caused by an open filter capacitor in the dc power supply results from feedback through the B+ stage from stage to stage. The filter capacitor not only removes power-supply ripple, but it also decouples those signals borne on the B+ line to keep them from being fed back.

The filter capacitor forms a low-impedance path to ground for signals, while remaining a high impedance for the dc B+ voltage. Because circuit resistances tend to be high, the frequency of oscillation caused by an open filter capacitor is low: "putt-putt-putt" is the sound of oscillations below 1 Hz.

The open grid circuit could be a tube with an open grid pin. I have been successful in "resuscitating" some of these tubes by simply soldering the grid pin to make the internal connection more reliable.

Another possible cause is an open IF transformer secondary winding. Still another is an open grid or grid-leak resistor. These resistors are especially likely to cause problems on antique radios because they tend to go bad with age. Older forms of resistors were not very stable. They aged to destruction both in service and in storage.

Noise Problems

Radio owners tend to refer to all forms of unwanted output from the receiver as "noise," but in this case we limit the form of noise to static noises. These noises can be divided into two classes: internal and external to the radio.

Some external static noises are naturally occurring (lightning flashes), and many are man made, such as noise from nearby power lines, automobile ignitions, or industrial machinery. In either case, since the problem is not inside the radio it is not discussed here. In this section our discussion will concentrate on noise generated inside the radio.

Diagnosing the noise as internal or external is relatively simple. Turn the radio on, listen to it, and note the noise. Next, short the antenna terminal(s) to ground and listen again. If the noise persists, then it is inside the radio. (*Note:* in some cases the perceived noise level will drop a little when this test is made, but the noise is nonetheless inside the radio.)

There are a number of different causes of internal noise problems in radio receivers, some of which follow:

- Noisy tubes (often they will arc internally, but this is seen only in subdued lighting)

- Poor, dirty, or loose tube socket connections
- Poor connections or solder joints
- Shorts in IF transformers
- Conductive dirt or foreign matter at crucial points (e.g., tube sockets or variable capacitor elements)
- Shorting tuning capacitors
- Shorting trimmer capacitors
- Arcs and shorts in the circuit

Diagnosis of the internal noise source is a case of divide and conquer. First, turn the volume control all the way down and note whether or not the noise has also disappeared. If the noise is still present, then the problem is between the volume control and the loudspeaker. If the noise disappears, then it is between the antenna and the top of the volume control.

Next, use signal tracing to locate which stage is bad. Once the bad stage is located, then the real fun begins: finding the bad component can be a bear. But here are some tips:

Suspected IF/RF transformer or plate load resistor   Remove the suspected tube from its socket. Connect a 10-kilohm, 2-watt resistor between the plate terminals and cathode terminals on the cathode of the tube socket. If the problem is in the plate load or IF/RF transformer, then the noise will persist. If it is the tube or the cathode resistor, then the noise will cease.

Suspected trimmer capacitor   The most common fault in this device is an electrical short between the plates (10-9A). This defect normally exists at points in the circuit where there is a high voltage across the trimmer capacitor, as might be found in certain local oscillator circuits.

Disconnect one side of the capacitor from the circuit and insert a 0.01 μF disk ceramic or mica capacitor in series with the capacitor (Fig. 10-9B). If the problem is the capacitor, then the noise will cease when the radio is turned back on and tested.

One of the most difficult types of noise to locate is internal arcing in fixed resistors and capacitors, and these problems occur frequently. There are three approaches to troubleshooting. First, tap the suspected components while listening for changes in the noise. This method is often unsatisfactory for two reasons:

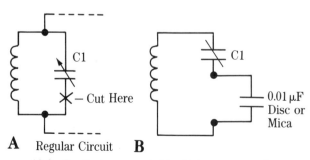

A   Regular Circuit   **B**

10-9   Testing a suspected bad trimmer capacitor.

tapping also vibrates nearby components, and some internal shorts are not vibration-sensitive.

The second method is replacement. But even temporary replacement takes time and does not appear professional unless it truly represents a cheaper and quicker way to do the job.

Finally, there is a piece of "test equipment" you can use to find the arcing component (Fig. 10-10). It is an ordinary medical stethoscope with the end bell removed. These instruments often come with home blood pressure kits, or are available from department stores that have a health goods department. Numerous mail-order sources also offer stethoscopes. In a pinch, you can also buy them from the same medical supplies outlets that your doctor and nurse friends buy their professional models from—but be prepared to pay a bundle. The tests we perform require only the cheapest grade stethoscope, so don't spend a lot of money unless you need the instrument for a medical reason as well.

Two cautions must be given: First, don't use a simple piece of hollow tubing with one end stuck in your ear. I did that at one time, but it turns out to be dangerous to the ear. Don't do it unless you want to visit the doctor after you fix the radio.

Second, use an ohmmeter on the highest resistance scale to determine whether or not the flexible tubing is actually an insulator. Some medical stethoscopes are

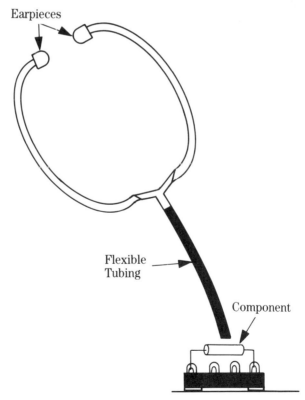

10-10   Testing for internal arcing. (See text for cautionary note.)

intended for use in operating rooms and other locations where flammable gases were once used in large quantities. Ether, cyclopropane, and other once-popular anesthesia gases were explosive.

In order to prevent static electricity from setting off an explosion, the equipment used in operating rooms was conductive and grounded. As a result, some stethoscopes were fitted with flexible tubing made from a carbonized rubberlike material—and it conducts electricity.

If you touch that rubber tube to a hot spot on the radio chassis, then you'll get an earfull of electrical shock, and no doubt will not speak highly of me for recommending this trick. Only perform this test after making absolutely sure that the stethoscope tubing is not an electrical conductor.

Use the stethoscope to probe around the bodies of suspected resistors, capacitors and other components listening for the telltale "tick-tick" sounds of an electrical arc. This method can be so sensitive that you can often tell which end of a capacitor is arcing, so go over each component carefully and thoroughly.

Oscillations

Oscillations, in the context of this section, are high-frequency tones that should not be in the receiver. Some squeals are an ordinary part of radio reception, especially at night when distant stations are being received. These signals are caused by heterodyning between two stations on adjacent channels. The offending station need not be audible.

I recall that many years ago, local listeners to a country music station on 730 KHz being bothered by a 10 KHz whistle. Complaints to the station brought an investigation by the chief engineer. He found that enough of the carrier of a 740 KHz station 110 miles away could be picked up to cause the squeal, even though the signal was audible only on the most sensitive receivers. A subsequent frequency change of one of the transmitters cured the problem.

External causes of squeals are found in the same way as external noise: short the antenna terminal(s) to ground and see if the noise disappears. However, this test has one limitation in the case of certain squeals that only show up on stations. The principal cause of this type of problem is an oscillating IF amplifier or IF/AVC circuit. The unmodulated IF oscillation is not heard unless a station is being received to heterodyne against it.

We can break the causes of oscillations and squeals into several categories: microphonics, chirps/birdies, tunable squeals, and other.

Microphonics

Microphonics are vibration-sensitive oscillations that start when the radio is jarred. The howling oscillation may dampen to zero rather quickly, or it may slowly build up to an unnerving "banshee" howl, or it may rapidly break into a squeal. The two principal causes of microphonics are tubes and tuning capacitors.

The tubes most often affected are RF amplifier and first audio, although other tubes do this trick as well. The cause is loose internal elements, especially suppressor grids in pentodes. These tubes generally test good in the tube tester, so they must be found by lightly tapping the envelope while monitoring the tube. Some tubes are so sensitive that tapping anywhere on the radio will cause the problem. In that case, replace the tubes with known good tubes one by one until the problem is found.

Tuning capacitors that cause microphonic oscillations are usually found to have one of several defects: loose tuning plates, bad rubber or fiber mounting grommets (which are used as shock absorbers), or dirt/dried grease under the rotor-grounding spring clip.

In the final case, carefully clean the surfaces underneath the ground clip and then clean the ball-bearing race(s) at the end(s) of the tuning shaft. Lightly relubricate the bearing race with a single small dab of white grease, such as Lubriplate®.

Chirps and birdies

Chirps and *birdies* are oscillations that occur only on one or on a small number of stations. There are several possible causes of these defects, but the principal among them are lack of neutralization in the RF amplifier, especially in TRF radios using triodes, image response in superheterodynes, and bad design. The final problem would probably exist only on earlier models or homemade radios because the art of selecting IF and LO frequencies to avoid problems has been well known since the 1920s.

The problem of neutralization comes from the fact that the tube's interelectrode capacitance provides a positive feedback path that will sustain oscillation at some or all frequencies within the radio's tuning range. If the neutralization capacitor is adjusted to a critical value, then the oscillation may break out either at a few points on the dial or only in one segment of the tuning range (typically the high end of the dial). Readjustment of the neutralization capacitor will correct the problem unless the capacitor itself is defective.

The problem of image response comes from the fact that the image frequency of the superheterodyne receiver gets through the mixer or converter stage. This problem is especially likely on small radios that don't have an RF amplifier stage. Recall from earlier chapters that the image frequency is located at the RF frequency $\pm\ 2 \times$ IF frequency, and is always on the other side of the local oscillator frequency from the RF frequency.

For example, in a standard AM band superhet with a 455 KHz IF and an LO on the high side of the RF dial frequency, the LO is at RF + 455 KHz, and the image will be found at the RF + 910 KHz. If the station being tuned is at 880 KHz, then the LO is at (880 + 455) KHz, or 1,335 KHz. The image frequency is [880 + (2 × 455)] KHz = [880 + 910] KHz = 1,790 KHz. The reason for this problem is that the 1790 KHz signal is also 455 KHz from the LO, so the difference frequency is 455 KHz—and is thus a valid signal as far as the IF amplifier is concerned.

Suppression of the image frequency should be done at the design stage. But in some cases, especially shortwave receivers, front-end tuning is so broad that images cannot be suppressed. The problem is especially noted in the tuning ranges above 10 MHz or so, when the IF frequency is 455 KHz or 460 KHz. Some manufacturers of car radios added image traps to the RF amplifier because car radios travel around to various areas, so they may be more susceptible even to out-of-band images.

Figure 10-11 shows a parallel-tuned image trap (L1C1) in the plate circuit of the RF amplifier stage. A parallel resonant LC tank circuit such as L1C1 offers high impedance to the resonant frequency, so it will prevent image signals from being passed from the RF amplifier to the mixer or converter stage. In order to make the image trap tunable, either the capacitor (C1) or the inductor (L1) must be mechanically ganged with the other tuned circuits in the radio. In many permeability-tuned car radios, one coil in the tuner assembly is an image trap.

The solution to the problem in other radios is to add either (or both) a *preselector* or a *wavetrap* to the antenna input circuitry of the RF amplifier.

Preselectors are used to sharpen the response of the input RF tuning circuits. The preselector adds additional tuned circuits, so it will narrow the bandwidth of the input RF-tuned circuits. If designed correctly, the preselector will attenuate the image frequency while passing the desired RF frequency.

Figure 10-12 shows two add-on preselector circuits. In both cases the preselector is passive; it contains no amplifier elements.

Active preselectors have at least one stage of RF amplification, but serve essentially the same purpose as the circuits of Fig. 10-12. These circuits must be built inside a well-shielded enclosure in order to prevent signals from simply bypassing the preselector by direct radiation to the antenna terminals of the radio—with the connection between the preselector and radio acting like an unofficial antenna!

Figure 10-13 shows a tuned wavetrap across the input circuit or antenna terminals of the radio. The wavetrap (L1C1) is a series-resonant tuned circuit, so it will present a low impedance to the resonant frequency and a high impedance to all other frequencies.

Because this series-resonant wavetrap is shunted across the signal path, the resonant frequency will be shunted to ground, while other frequencies are passed on to the antenna input circuits of the radio. The wavetrap is tuned to the image frequency.

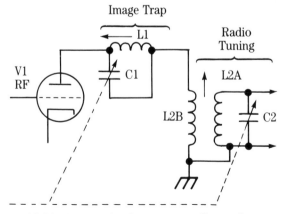

10-11   Image rejection trap on radio receiver.

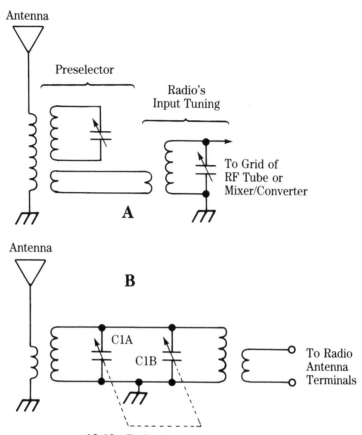

10-12   Radio input tuning methods.

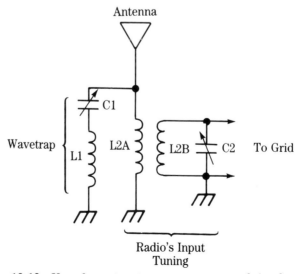

10-13   Use of wavetrap to suppress unwanted signals.

The wavetrap is also used to attenuate large in-band signals that tend to overload the receiver. If the radio is close to a broadcast station it may tend to overload the receiver input circuits, resulting in intermodulation products or desensitization of the radio. By tuning the wavetrap to the offending station one regains control over the problem.

Tunable oscillations

Tunable oscillations are squeals found at many points across the band. They may appear on all stations, but they may only exist when the radio is tuned to a station. The important attribute of these squeals is that they show up at different discrete points on the dial instead of being either isolated or all over the dial and not subject to tuning, such as when the audio amplifier oscillates.

Tunable squeals can be caused by a number of problems. Look for an open AVC bypass capacitor, especially if the radio distorts on strong local stations but not on weak distant stations. Otherwise, look for loose tube shields, especially in the RF and IF amplifiers, broken ground connections, or open bypass capacitors (especially in the plate load screen-grid circuits).

In the case of the tube shields, it might be a good idea to clean both the shield and its contact point on the socket as a matter of course, especially when oscillations are noted. Do not depend on an ohmmeter to find the problem because the tube will appear to be grounded for dc but might be floating a small potential above ground for RF. If a tube shield is missing, then replace it. Shielded sockets are obvious in most cases, so you can easily tell whether or not a shield is absent.

Bypass capacitors can be tested by temporarily bridging a 0.1 µF/600 WVdc capacitor across the suspect capacitors, each in its turn. The oscillation will cease when the defective capacitor is found.

Other causes of tunable oscillation are also found sometimes. For example, a tube with an open suppressor grid might check good in the tube tester, but oscillate in the circuit. Testing by replacement is the only option. Also, open filter capacitors may cause loss of decoupling (hence oscillation) even though it has little effect on hum due to power-supply ripple.

Some radios have a paper capacitor, typically 0.01 µF to 0.5 µF, bridged across the B+ line and ground, in parallel with the filter capacitor, to account for the fact that electrolytic capacitors are less effective at RF frequencies. These capacitors may be open.

The oscillation also might be caused by misdressed wiring from previous service attempts. If wiring is dressed so that output circuits in a stage are in close proximity to the input circuits, then coupling and therefore feedback may result. Redressing the wires to another location will solve the problem.

On older radios, such as those serviced by many readers of this book, the oscillation may be due to surface dirt on the chassis and components. The dirt and built-up grease can form a high-impedance path for RF (indeed for dc too!) that allows feedback to form between stages and between the input and output of single stages. Cleaning with an appropriate solvent will work wonders.

I recall one ham radio receiver (an expensive model) that had been used in the kitchen of a ham's house (don't laugh, I've operated my station from all rooms of the house!). Airborne grease from cooking coated the cabinet, chassis, and components. Cleaning the radio made it perform much better.

Other high-frequency oscillations

We've already covered most forms of oscillation, but there are still at least two forms. First, there is the possibility of dual-mode oscillation of the IF amplifier. If the input and output tuned circuits of the IF amplifier are set to slightly different frequencies, and other problems exist (see previous sections of this chapter) then it is possible for an IF amplifier to oscillate on two different frequencies. If these frequencies are close together, they will heterodyne together and cause a squeal.

The second type of oscillation is oscillation of the audio amplifier stages. Suspect the audio circuits if the frequency of oscillation is affected by either the volume or tone controls, or if the oscillation still exists when the volume control is turned down.

The causes of oscillation in audio stages are similar to those for RF/IF stages, so they will not be covered separately here. In addition, however, look for open grid resistors in the audio amplifier stages.

If a loudspeaker has just been replaced, and if the radio employs audio feedback to improve linearity (to reduce distortion), then try reversing the leads between the voice coil and the audio output transformer secondary. If these connections are not made correctly, then either the feedback network will be grounded and therefore eliminated, or a reverse phase signal will be fed back and will cause oscillation.

Distortion

A radio should have a rich, mellow tone when tuned to a good station. But when nonlinearity creeps into the circuits, the radio will emit a harsh distorted sound. In this section we will examine some of the common causes of distortion, and the troubleshooting methods needed to find them.

Note whether or not the distortion exists on all stations or just on strong, local stations. If weak stations are clear, but local stations are distorted, then suspect either the RF, IF, converter/mixer, or AVC circuits.

In the case of the RF, IF, and converter/mixer stages, the defect might be a shorted tube or an improper bias. Check cathode resistors and bypass capacitors. In transistor radios that employ an RF amplifier stage, the RF transistor might have a high leakage resistance between the collector and base and still test good on a transistor tester.

On all radios check the bypass capacitors in the AVC for either opens, shorts, or leaks (high-resistance shorts). Also check the AVC resistors for change resistance value. Now let's look at other causes of distortion.

Loudspeaker

If the loudspeaker voice coil is jammed, if the cone is badly torn or warped, or if the speaker frame is warped, then the sound produced by the speaker will be distorted. Replace the loudspeaker with a new speaker, or recone the old one if a new model is not easily available.

Defective vacuum tubes

If the vacuum tubes in the audio stages are shorted, then they will be nonlinear and will produce a distorted sound. Check the vacuum tubes carefully, or test by temporary replacement.

Shorted cathode bypass capacitors

If the bypass capacitors across the cathode resistors in the audio amplifiers are shorted, then the cathode resistor is also shorted to ground through the capacitor. This fault will set the bias on the tube served by the capacitor to an incorrect point, resulting in distortion.

Open or increased-value resistors

Resistors are not as stable as one might think, so they are often found with a much higher-than-marked value. Check resistors with an ohmmeter. In some circuits the resistor will have to be unsoldered from the circuit at one end so that circuit resistances don't obscure the reading.

Look especially at cathode bias resistors, plate load resistors, control grid resistors, and screen grid voltage-dropping resistors. If a feedback network is used, make sure that the resistors and capacitors in the feedback network are good.

Shorted or leaky coupling capacitors

Most radios use resistance-capacitance coupling between stages in the audio amplifier. If the capacitor is shorted, or has a high-resistance "leakage" short, then the B+ from the previous stage will appear on the normally negative grid of the following stage (Fig. 10-14).

The audio output tube may appear to be shorted because the plate may glow red. Although many tubes can survive a shorted input coupling capacitor, one must always test the tubes after this fault is found to determine whether any secondary damage occurred.

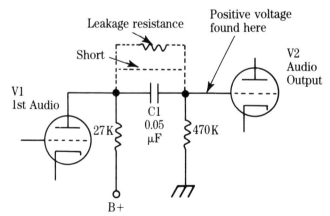

10-14   Effect of shorted or leaky capacitors is to place high voltage from V1 anode onto V2 grid.

Shorted or open output-transformer winding

If the primary or the secondary winding of the audio output-transformer has a partial short—only part of the winding is still viable—then the resulting impedance mismatch will cause distortion.

On push-pull amplifiers, one side of the primary winding may be open, resulting in an unbalanced push-pull amplifier producing severe distortion. This fault is easy to find because the B+ voltage will be missing on one of the push-pull vacuum tubes, but not the other.

Some signal tracers, radio analysts, and other test instruments have a built-in output transformer that can be used to substitute for the suspected bad transformer in the radio.

11
Troubleshooting
Intermittent Problems

REPAIR PROBLEMS ARE NEVER FUN, BUT SOME PROBLEMS ARE WORSE THAN OTHERS. Probably the worst of all are the dreaded *intermittents*. These problems come and go, usually in the least convenient manner. They almost never occur when you are ready to troubleshoot. You will stand there, instrument probes in hand, brain engaged and on the alert, only to find that darn radio set working perfectly. Turn your back and—zot!—the trouble appears momentarily and then heals itself again.

What is one to do? Finding intermittents often falls into the same areas as sorcery and witchcraft, but there are certain things that can be done to enhance the probability of success.

One thing you can do is educate yourself regarding the kinds of parts and faults that are most likely to create intermittent symptoms. Very high on the list are switches and relays. These devices are mechanical, so they are subject to wear and tear. You will find dirty internal electrical contacts, poor spring tension and other faults.

In many cases, a session with a pencil eraser on the contacts or a squirt of contact cleaner (such as *Blue Stuff*®) will work wonders. In other cases, only replacement of the part will solve the problem.

Another sore spot is potentiometers used for volume and tone controls. These components are variable resistors in which a shaft-operated wiper electrode rubs against a wirewound or carbon resistance element. If either the element or electrical contact on the wiper gets dirty, then operation can become intermittent. Unless the dirt has physically damaged the resistance element, as sometimes happens, especially on carbon elements, a simple squirt of contact cleaner will solve the problem.

Be especially wary of potentiometers that normally pass dc through the wiper connection. I recall a 1963 car radio model where cost-conscious engineers eliminated a coupling capacitor from the volume control and audio preamplifier circuit, thereby making the volume control resistance part of the preamp transistor's dc bias network. Passing the dc bias current through the control generated a massive warranty problem for the manufacturer. Those volume controls were chewed up by the truckload!

The printed wiring board (PWB) is another common source of intermittent problems. Two problems are common: poor solder joints and damage to the board, which is sometimes hidden. Both types of fault are aggravated by heat: near power transistors, rectifiers, vacuum tubes, power resistors (2 watts and above), lamps, and so forth. We will discuss PWB problems in a moment.

In vacuum tube equipment, the tube socket can produce intermittent problems. If the tube pins or the socket contact loses tension, an intermittent connection results (Fig. 11-1). These faults can be repaired in most cases. If dirt is the problem, then remove the tube and gently clean its pins with a dime store ink eraser, spray the pins with a clear contact cleaner and reinsert the tube into the socket.

Next, pull the tube out of the socket and reinsert it four or five times. This action will clean the socket. Wait a half hour or so for the cleaner to dry, and then turn the radio on to evaluate the results. If the intermittent remains, then re-tension the socket contacts with either a tiny screwdriver tip or a sharp pointed tool.

Components can be the source of some maddening intermittents. Unfortunately, some component problems tend to heal themselves the instant a probe is attached. Be especially wary of plastic packaged transistors, tubular (non-Mylar) capacitors and resistors. These components account for a large portion of the problems.

Finding the Intermittent

There is no such thing as a universal procedure for locating intermittents because different equipment has somewhat different requirements. The first step is to observe the problem. Define what the set is doing wrong, what functions are affected and, if the set is a transceiver (ham or CB), whether it happens on both receive and transmit.

This step is crucial to quick success. You can, after all, restrict your efforts to a few stages. For example, if the transceiver problem happens on SSB but not CW, or

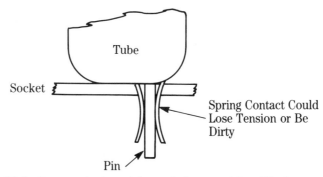

11-1 Lose contacts on tube sockets cause intermittents.

on receive but not on transmit, then we can infer that the problem is probably not in a stage that is common to both affected operating modes. Determining which stages are likely candidates is done by understanding the affected set. Do a block diagram analysis.

Mechanically intermittent

Some intermittents occur while being vibrated, touched or thumped. A certain number of such problems are due to bad switches and potentiometers. A little light tapping with an insulated probe, jiggling of the device, or visual inspection will find the problem.

More mechanical faults are caused by bad PWB solder joints or bad components that are mounted on the PWB. Again, light tapping on the board or components will often yield the problem. Unfortunately, many such problems fail to yield to the tapping method because any vibration at all, anywhere on the board, will produce the fault. There are two approaches that are helpful in this case, which do not require formal troubleshooting: visual inspection and shotgun solder touchup.

Visual inspection involves looking at every joint on the PWB. It helps to use a 10× or so magnifying glass and adequate light. Examine the board two ways. First shine the light on the soldered side, second, shine the light through the PWB from the component side (Fig. 11-2).

In the latter case, subsurface cracks in the PWB material can break a joint or track. Even if the joint or track appears normal it should be reworked. Visual examination takes a certain amount of practice. One needs to develop a "small eye;" that is, the ability to see defects where unskilled people see a "normal" joint.

I usually inspect PWBs with a bottle of fingernail polish or a grease pencil handy. Each apparent anomaly is marked so that I can find it later on. This habit is especially useful when using a magnifier because the glass will distort space perceptions.

Shotgun soldering is used especially when the area of the intermittent is known, the PWB is small, or when nothing else seems to work. I can recall an FM car radio receiver problem in which the VHF FM front-end PWB was difficult to remove and

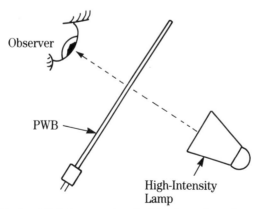

11-2   Use backlighting on printed wiring boards to detect faults.

replace, yet these boards had a high "bad joint" intermittent rate. In that case, the more elegant visual inspection method was not practical, so we just pulled the PWB, soldered every joint and tinned every track. Rarely did this method fail on that particular problem.

I admit that the elegant method is to find the single bad joint or broken track, and repair only it. Unfortunately, that can be time consuming and sometimes impossible. While the purist "supertech" is trying to analyze which joint is bad, I am going to fix the set! The practical approach is more profitable.

Thermal intermittents

Another category of intermittent is sensitive to temperature differences; both hot and cold defects are seen. The most blatant form is the set that refuses to operate properly under either hot or cold. During the winter the temperatures reached by automobile equipment is the local overnight temperature—say 40°F in northern states or Canada.

During the summer, on the other hand, temperatures of automobile equipment will be considerably above local air temperatures. In 1963, when a major automobile electronics company began experiencing reliability problems with their new solid-state radios, they asked employees to leave their cars unlocked so that the engineers could measure the cabin temperatures. After four hours of 90°F sun, the interior temperatures were found to be 140°F at the front seat and up to 180°F behind the dashboard.

Extremes of temperature seen by car radios results in some peculiar intermittents. One form is the set that won't work when you get in the car, but it does work ten minutes later after the heater or air conditioner has altered cabin temperatures.

Radios used in the home do not experience the extremes of ambient temperature seen by car radios, but nonetheless they experience related intermittents. Typically, such a set will either fail to work at all until it heats up, or it works nicely until it reaches temperature and then fails. Even when an intermittent is not specifically related to temperature, it is frequently true that temperature aggravates the situation—allowing you to find it easier.

First, let's talk about how to heat up a set. Use general area heating only for determining that the fault is temperature sensitive. A high-wattage lamp, sun lamp or hair dryer can be used. In equipment where there are a lot of power devices (or vacuum tubes), we can often heat up the circuits by placing a towel or blanket over it. This method is particularly useful for thermal faults that occur only in the cabinet. The thermal fault will continue to exist for a few minutes after the towel is removed, allowing troubleshooting time.

Area heating will give you the time needed to troubleshoot the fault. It will not, however, help you find a specific thermally sensitive component. For this chore we must use local area heating.

Several methods are available for local heating. A small "tensor" lamp, for example, will allow heating of a small area on a PWB. A soldering iron or gun will concentrate heat very locally, almost to the exact component. Be careful, however, because the hot tip of the soldering tool will damage some components, especially polyethylene capacitors.

Another method, used for heating individual components, is shown in Fig. 11-3. The heat source is a 6- or 12-volt incandescent lamp, such as a number 47 or number 1891. A small cylinder made of some material, such as insulated sleeving ("spaghetti") is designed to fit over components such as transistors and some integrated circuits. The heat source is placed in the open end, thereby concentrating the heat only on the component under suspicion. The tale will be told in about 30 seconds.

The indication that the component being tested is bad will be obvious. There will be a sudden change of operation, or a sudden increase in the noise produced by the circuit. Rarely is the change subtle.

The other thermal problem is that equipment fails to operate when cold. In this context, cold could mean arctic temperatures, room temperatures, or anything in between.

Area cooling is harder to do, in some cases, than area heating. But for small devices (up to, say, the size of a table model radio) try putting the unit in the refrigerator for about an hour. I still fondly recall the shocked looks on the face of boat radio customers when I placed their "won't work on cold days" VHF-FM transceiver in the fridge. Many of those rigs are merely overpriced variants of 2-meter FM ham rigs. In most cases, 30 to 60 minutes in the refrigerator yields 5 to 10 minutes of troubleshooting time.

Local cooling is needed to isolate components. For this case use a can of freon "freeze spray" (Fig. 11-4). Electronic supply stores sell this product under several

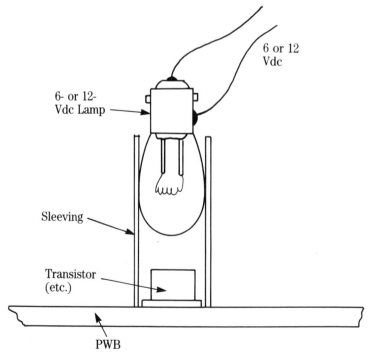

11-3   A localized heat method for finding thermal intermittents.

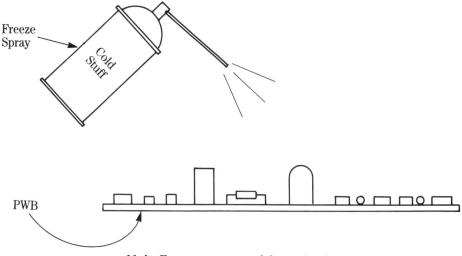

Freeze
Spray

Cold
Stuff

PWB

11-4 Freeze spray to cool down circuits.

brand names. The types of stores most likely to have freeze spray are those whose clientele includes many radio-TV-audio repair shops.

Be careful not to spray too much area. Freeze spray is expensive and general area cooling will not locate the bad component. Use the spray only on single suspect components or small groups of components.

You can verify the bad component by reheating it with a soldering iron or the gizmo shown in Fig. 11-3. If the problem repeatedly appears and disappears on heating/cooling cycles, then you have found the problem. Even if the problem is not consistently repeatable, however, we can "work the odds" and replace the component on speculation.

Our final method for fixing intermittents is another shotgun approach. In this case, however, we replace components on a "scattergun" basis. I admit it's not very elegant, and provides no balm at all to salve the ego of the technical genius. After all, anyone can unsolder a half dozen components and replace them.

But let's consider some facts. I once worked in a hospital electronics laboratory that repaired clinical equipment. The emphasis was on low-cost, rapid repairs. One famous brand-name patient monitor used vintage circuitry. The ECG preamplifier and the dc power-supply regulator used literally dozens of 2N3393, 2N3906 and 2N3904 plastic small-signal transistors. These transistors were typically connected six to eight at a time in circuits with multiple feedback and signal paths—all direct coupled.

Troubleshooting is time-consuming in such circuits. At that time, those transistors cost us $25/100 in bulk packed bags. It takes 15 minutes to replace eight small transistors (that cost us $2). A total of 30 minutes puts the equipment back on line.

The situation is only a little different for you. The biggest difference is that you buy transistors in overpriced blister packs rather than 100 lots. Nevertheless, when the troubleshooting problem seems intractable, shotgunning components is a viable alternative.

Problems with IF and RF Transformers

The IF and RF transformer represents a high potential for intermittent problems in radio receivers. There are two basic forms of problems with the IF transformer: intermittent operation and intermittent noise. The best cure for a bad IF or RF transformer is replacement, but because old IF and RF transformers are not always available today, we must place more emphasis on repair of the transformer.

Figure 11-5A shows the basic circuit for a single-tuned IF transformer (others might have a tuned secondary winding). If one of the very fine wires making up the coil breaks (Fig. 11-5B), then operation is interrupted. The problem can usually be diagnosed by light tapping on the shield can of the transformer, or by using a signal tracer or signal generator to find where the signal is interrupted. In some cases, the plate voltage of the IF amplifier is interrupted when the transformer opens, and this can be spotted using a dc voltmeter.

Repairing the IF transformer is a delicate operation. Examine the shield can to determine how it is assembled. If it is secured with a screw or nut, then merely remove the screw or nut. If the IF transformer is sealed by metal tabs, then you must very carefully pry the tabs open (don't break them, or bend them too far!). Slide the base and coil assembly out of the shield. If there is enough slack on the broken wire, then simply solder the wire back onto the terminal.

The heat of the soldering iron tip will burn away the enamel insulation. If the wire does not have enough slack, then add a little length to the terminal by soldering a short piece of solid wire to the terminal, and then soldering the IF transformer wire to the added wire. In some cases, it is possible to remove a portion of one turn of the transformer winding in order to gain extra length, but this procedure will change the tuning.

Most noisy IF transformers are bad because the capacitor inside the shielded can is intermittently shorted. Figure 11-6 shows a method for locating the noise source. First, isolate the stage using the signal tracer or signal injection method.

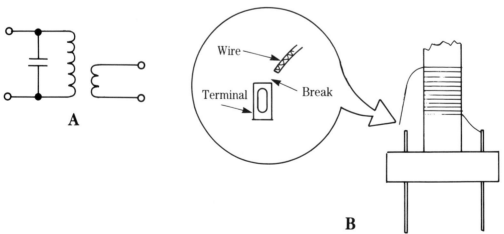

11-5 (A) IF/RF transformer or coil. (B) Typical failure mechanism.

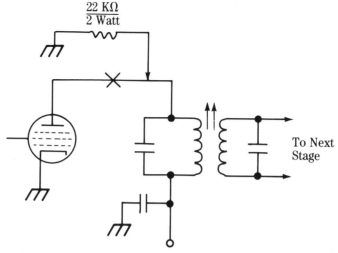

11-6   Use a substitute "plate load" to isolate noisy IF transformer.

Then turn off the set and disconnect the ac power cord. On ac-only sets, without series wired filaments, it is simple to remove the tube from its socket to make this test.

In other sets with series strung filaments it is necessary to disconnect the plate lead between the IF transformer and the amplifier tube, as shown in Fig. 11-6. In either case, connect a 22-kilohm, 2-watt resistor between the IF transformer plate lead and ground. Turn the set back on and let it warm up. If the noise is in that transformer, then it will reappear.

The noisy IF transformer needs replacement more than open types, but we are still faced with the lack of original or replacement components. The capacitor can be

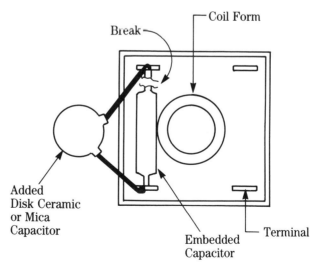

11-7   Using a disk ceramic or silvered mica capacitor to repair unavailable IF transformer.

replaced on many IF transformers, and that can lead to a repair. If the capacitor is a discrete component soldered to the base, then it is simple to replace it with another similar, if not identical capacitor. But if the IF transformer uses an embedded mica capacitor (Fig. 11-7), then the job becomes more complex.

If the capacitor element is visible, as in Fig. 11-7, then clip one end of the capacitor where it is attached to the terminal. Then bridge a disk ceramic or mica capacitor between the two terminals. The value of the capacitor must be found experimentally if the value of the coil is unknown. It is the value that will resonate the coil to the IF frequency.

If the capacitor is embedded and not visible, then it may become necessary to install a new terminal and solder the added capacitor to that terminal, with appropriate external circuit modifications.

12
Radio Alignment Techniques

OF ALL THE JOBS PERFORMED BY THE RADIO SERVICE TECHNICIAN, NONE IS MORE "technical" or requires more skill than radio alignment. Many antique and classic radio collectors who do much of their own work most of the time will shy away from the alignment task, and refer it instead to specialists. In the next few pages, the basic principles and practices of radio alignment will be demonstrated, and the tools required to perform the alignment job will be described.

AM Radio Receivers

The simple AM radio is the earliest to align. The test equipment is both simpler and less costly than that needed for either FM or stereo multiplex alignment. A service-grade RF signal generator covering the range of 400 to 1600 KHz (200 KHz to 1600 KHz for car radios) creates the necessary controlled local test signal. These signal generators are relatively low cost when new, and can often be purchased second-hand for very little and restored.

A good alternative is a crystal oscillator, featuring switch-selectable outputs at 455 KHz (262.5 KHz for car radios), and 500 KHz. If a large number of European radios are to be serviced, then the technician should have a unit with 460 KHz available. Specified frequencies are 262.5 KHz for the standard U.S. car radio IF, 455 KHz for the IF in all home radios and some older car radios, 460 KHz for the standard European IF, and 500 KHz for a dial point marker.

The harmonics of the 500 KHz crystal will identify the low end of the AM band, the midpoint at 1000 KHz, and a high point of 1500 KHz. It is also possible to use a 1000 KHz crystal, feeding a chain of IC frequency dividers to gain the required markers. A single J-K flip-flop will divide by two and produce 500 KHz. A decade divider IC or four J-K flip-flops connected as a mod-10 divider will give you 100 KHz check points.

Coupling the signal generator to the radio

Most RF signal generators have a 50-ohm output impedance. In most brands of signal generators, this figure is resistive at low frequencies, so impedance matching is not usually needed. In many cases, however, some form of dummy antenna or impedance-matching pad is needed between the signal generator and the radio antenna input terminal.

Most car radios, for example, use an antenna input impedance higher than 50 ohms (120 ohms is common). Because of this, it is necessary to simulate both the input impedance of the radio and the capacitive effects of the antenna feedline. Although the antenna cable has a lower capacitance than normally used in radio work, it must still be considered.

This capacitance is shunted across the capacitor that tunes the input tank circuit in the radio. To compensate for the different capacitance values and to match the generator to the radio, you generally should use a dummy antenna load, such as that illustrated in Fig. 12-1A. The value of the shunt capacitor is usually 30 pF.

The series capacitor, however, tends to vary from brand to brand. In some it is specified as either 30 pF or 60 pF; others require more unusual values. The particular value required for any specific model car radio will be found in either the manufacturer's service manuals or the AR-series *Photofact* books.

Delco recommends using the dummy load shown in Fig. 12-1B for its solid-state radios. Although this coupling method will allow only a rough adjustment of the car radio antenna trimmer-capacitor, it is best to use the manufacturer's specification on Delco radios. All of these dummy loads can be built into the standard Motorola plugs universally used as the car radio antenna connector.

Figure 12-2 shows the proper methods for connecting the signal generator directly to home radio circuitry to align a seriously out-of-alignment IF amplifier stage. Initially connect the signal-generator output through a 0.1 μF/600 WVdc capacitor to the grid of the IF amplifier tube. This connection is used to align the output IF transformer tuned circuits when they are so far out of alignment that the signal will not get through the stage. However, this connection will not allow alignment of the input IF transformer.

When the output IF transformer is aligned, or if it was never too far out of alignment to pass at least some signal, use the alternate method shown in Fig. 12-2: connect the generator to the RF input grid on the mixer or converter stage. The output of this stage is connected to the input of the IF amplifier, so it is a good point

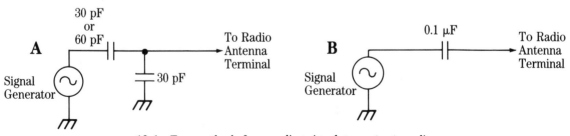

12-1   Two methods for coupling signal generator to radio.

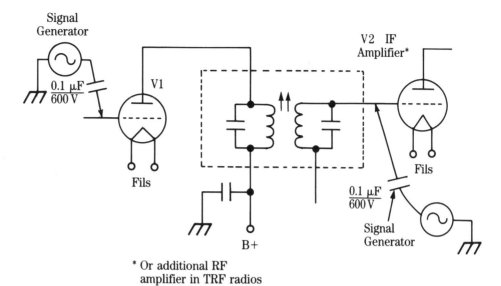

Signal
Generator

$\frac{0.1\ \mu F}{600\ V}$

V1

V2   IF
Amplifier*

Fils

Fils

$\frac{0.1\ \mu F}{600\ V}$

Signal
Generator

B+

* Or additional RF
amplifier in TRF radios

12-2   Injection point for IF signal.

to inject the IF frequency test signal, and it is isolated from both input and output IF transformers.

Coupling to the RF amplifier input stage can be done in either of two ways, depending upon the design of the radio. If the radio uses an external antenna, then connect the signal generator to the circuit through a 0.1 μF capacitor and the antenna terminals of the radio. In radios that have a loop or loopstick antenna, connect the output of the signal generator to a homemade loop of hook-up wire. Use six to ten turns of No. 22 wire in a circle about 2 inches in diameter. Closely couple the signal generator to the loop antenna.

You may also use a "gimmick" to couple signal to the loop antenna. The "gimmick" in this case consists of about 25 turns of wire closely wound on a 3/8-inch form. The wire is not grounded, but it is connected to the signal generator output.

Another "gimmick" injection method is shown in Fig. 12-3C. In this case, we are injecting the signal into a shielded RF coil or IF transformer. The trick is a 1-inch to 3-inch piece of insulated hookup wire, with one end stripped of the insulation. The wire is connected to the output of the signal generator, and is then dropped into the open end of the tuning core inside the RF or IF coil. This method works best with the type of core that has a hexagonal hole rather than a screw slot.

Signal output-level indicator

The second piece of equipment needed for alignment is some instrument to measure relative signal output level. This instrument can be either an oscilloscope or ac VTVM across the speaker terminals (Fig. 12-4). If this method is chosen, however, it is necessary to use a modulated signal generator so that an audio output signal is produced. An unmodulated signal generator will not produce an audio signal output.

Also, when using this system it might be wise to load the output of the radio with a power resistor, rather than a speaker. Alignment can be rather nerve racking,

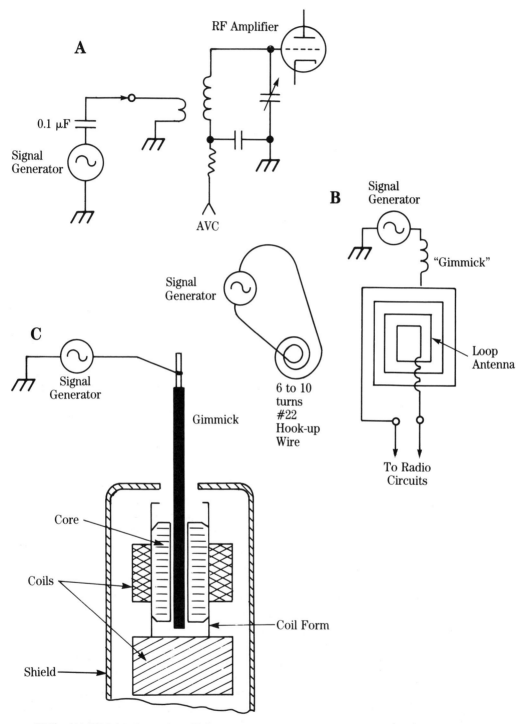

12-3 (A) RF injection point. (B) Loop antenna signal injection. (C) Use of a "gimmick" to inject signal.

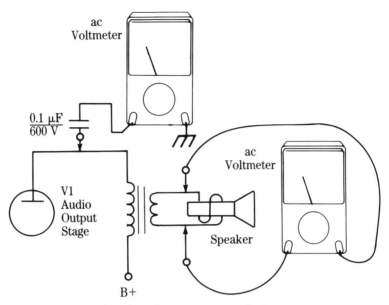

12-4 Ac voltmeter as output indicator.

especially if several technicians are working close together, because of the shrill tone from the radio speaker. The silent approach to alignment will assist in keeping peace in the shop. A proper resistor to substitute for the speaker on most radios would have a value of 4 or 8 ohms at 20 watts or 10 ohms for certain later Delco car radio receivers.

Alternatively, an ac voltmeter or oscilloscope can be connected to the plate of the audio power-amplifier tube. Because there is a high-voltage B+ potential at this point, it is necessary to use a 0.1 μF/600 WVdc capacitor between the meter and the circuit.

Another useful output signal level indicator is a dc VTVM connected across points affected by the automatic gain control circuit or automatic volume control of the radio (Fig. 12-5). The AGC circuit line provides a bias voltage to the RF amplifier, the IF amplifier, and sometimes to the converter circuit, in proportion to the signal level. When the signal level is high, the bias forces the amplifier gains lower, and when the signal level is low gain is forced higher.

The result is a relatively constant signal output level from the IF amplifier, and a bias voltage that is proportional to that signal level. In tube radios (as shown in Fig. 12-5) the AGC or AVC line has a negative bias voltage on it, so connect the positive terminal of the dc voltmeter to ground or common and the negative terminal of the meter to the AVC/AGC line.

In transistor radios using an npn RF stage, connect the VTVM from either the AGC control line to ground or the emitter of the RF amplifier transistor to ground. In transistor radios using pnp transistors, connect the VTVM from the RF collector to ground. Or, connect the negative VTVM lead to the radio B+ line and the positive VTVM lead to the RF transistor emitter. The collector connection is both easier and more acceptable.

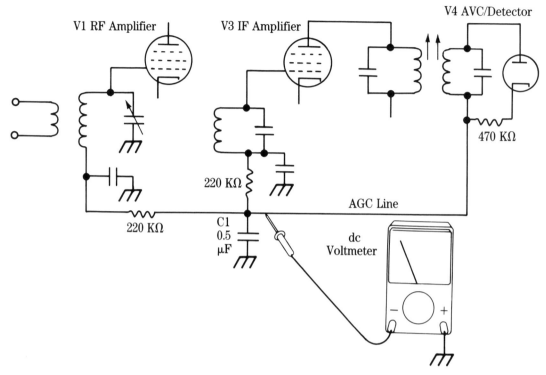

12-5   Dc voltmeter on AGC line as output indicator.

In any event, whatever output indicator is used, it is necessary to keep the RF output from the signal generator low enough to prevent the AGC from upsetting your indications. Up to a certain signal level, the AGC causes a linear decrease in the gain of the RF/IF amplifiers, proportional to the increase in signal amplitude.

Beyond that critical signal level, however, the radio will not reduce the gain in proportion to the signal level. This is called the *AGC saturation point*, or the *AGC knee*. Operating the signal generator at a level above the AGC knee will result in ambiguous readings, and poor alignment.

In summary, Fig. 12-6 shows the proper hookup of test equipment for aligning a typical AM radio. We have discussed the proper connection of the instruments. A power-line voltage meter is an optional but useful accessory. It is wise to ensure that this voltage is set for the value specified (115 Vac). This problem is not too severe on AM radios and modern FM radios, but on classic FM radios it is worse. Also, in auto radios, standard voltages vary among cars, but will be indicated in the service literature.

AM alignment procedure

It is necessary to begin the alignment with the IF amplifier. While remembering to keep the RF level just high enough to get a reasonable indication, peak the tuning cores of the IF transformers. Start with the secondary tuning core on the output IF transistor. The secondary is usually the bottom core. Next, align the primary core of

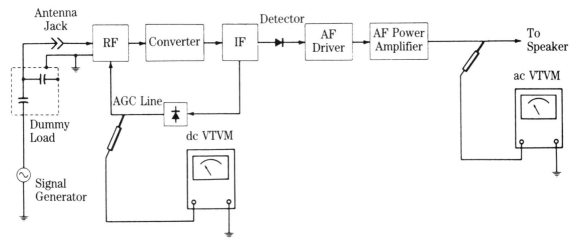

12-6 Ac and dc methods for signal level indicator.

the same transformer. Proceed next to the secondary core of the input IF transformer.

After peaking all of the IF transformers, go back and repeat the process several times until no further increase in signal level can be produced. This is necessary because there is a certain amount of interaction between the various tuned circuits. If the radio is badly out of alignment, it will be necessary to reduce the generator output several times during the course of alignment.

The front end of the radio is aligned by feeding the signal into the antenna circuit. The first step is to set the oscillator tracking. Pick a generator frequency of known accuracy that corresponds to a frequency marking on the dial of the radio (1500 KHz is a good one on most radios).

If the accuracy of your signal source is unknown, try using a broadcast station signal for this adjustment. The FCC requires AM stations to maintain their frequency to within 20 Hz of the channel assignment. When both the radio and the generator are dialed into the same frequency, adjust the oscillator trimmer-capacitor for maximum signal as read by your level meter. This sets the oscillator frequency so that it tracks with the dial frequencies.

If the tracking changes between the low end of the band and the high end, it will be necessary to adjust the LC ratio between the oscillator capacitance and inductor. The standard procedure is to adjust the trimmer capacitor for high end of the band (1500 KHz) tracking and the coil slug for low end of the band (600 KHz) tracking. The final step is to peak the RF amplifier collector and (if used) antenna tuning capacitors for maximum output. These steps must be repeated several times to insure accuracy.

In car radios antenna peaking must also be repeated when the radio is remounted in the car and whenever changes are made to the antenna system of the car. One purpose for this trimmer is to compensate for variations between one antenna and another. If it is not performed in the car, the radio will remain tuned either to your bench antenna or your dummy load, with the usual decrease in performance from the user's point of view.

The trimmer adjustment is frequently omitted by mechanics who install radios. If a newly installed or reinstalled car radio is presented for repair, and it appears to have either antenna or RF amplifier troubles, try peaking the trimmer before removing the radio from the car. This adjustment is made on a weak station above 1400 KHz. The trimmer-capacitor adjustment will be found either on the chassis beside the antenna jack or behind both the front panel and control knobs.

FM Radio Receivers

FM alignment can be performed in one of two ways. One method uses an unmodulated generator and the other uses a sweep generator. Of the two methods, the sweep generator procedure is preferred if suitable test equipment is at hand. A cheap sweep generator, however, can negate your efforts enough to make the unswept method more successful.

Alignment tools

The dummy antennas shown in Fig. 12-7 are usually recommended by various radio manufacturers. Most FM receivers use either a 75-ohm unbalanced coaxial cable or 300-ohm balanced twinlead antenna input circuit. The two-resistor dummy antenna in Fig. 12-7A offers a better match between the 50-ohm output that is common on RF signal generators and 75-ohm radios. On 300-ohm radios use the pad of Fig. 12-7A on one antenna terminal, and ground the other. Alternatively, use a 4:1 ratio "Balun" transformer, available from video suppliers—TV systems use the same hardware. This transformer will convert the 75-ohm output impedance from the resistor pad to 300 ohms required by the radio.

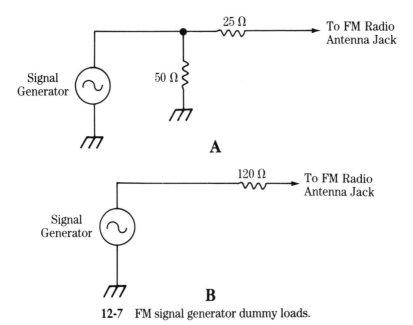

12-7 FM signal generator dummy loads.

Some car radio manufacturers use the version shown in Fig. 12-7B. This pad reflects the fact that the car radio antenna is typically 120 ohms. The capacitive dummy antenna used on AM alignment will attenuate the signal too much and detune the FM radio input.

Also needed in FM alignment, especially if sweep generator procedures are being used, are suitable oscilloscope and VTVM probes. One primary probe used in many applications, including alignment, is the low-capacitance probe. This style probe, illustrated in Fig. 12-8, consists of a 10-megohm resistor shunted by a 30 pF trimmer-capacitor. The trimmer is adjusted for optimum performance by feeding a square wave through the probe to an oscilloscope.

In most applications a 1000-hertz square wave is specified. The trimmer is adjusted until the squarest waveform is presented on the screen of the oscilloscope. The probe features a 10:1 attenuation ratio, which means the scope will display one-tenth the actual voltage applied to the probe. This is important to keep in mind when using such a probe. This probe can either be homemade, or it can be purchased for under $10 in kit form.

Figure 12-9 shows two demodulator probes used in FM alignment. The passive probe of Fig. 12-9A is the usual type employed. This probe also has an attenuation factor. The one redeeming quality of this style probe is that it is both effective and low in cost. The more complex probe of Fig. 12-9B is becoming increasingly popular. It does away with the attenuation factor and in fact offers amplification.

Specific circuits for this style of probe can be found in the service literature of appropriate radio manufacturers. The transistor amplifiers A1 and A2 are capacitively coupled. Simple stages using either bipolar or junction field-effect transistors (JFETs) can be employed provided they have a wide frequency response. These probes must respond to at least the FM IF frequency of 10.7 MHz. It is desirable that the probe also provide at least some gain greater than unity up to the FM local oscillator maximum frequency of 120 MHz. This is a desirable option but is not mandatory in most instances, since the majority of all measurements will be made in the FM IF amplifier stages.

Alignment using an unmodulated generator

Figure 12-10 shows the basic equipment hookup for performing an FM alignment using an unmodulated signal source. The signal generator can be a simple service-

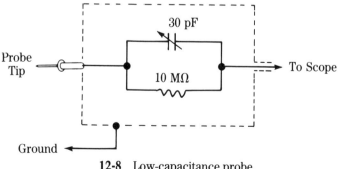

12-8 Low-capacitance probe.

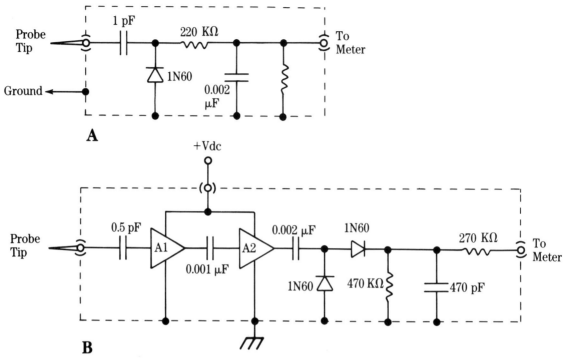

12-9   RF demodulator probes: (A) passive, (B) active.

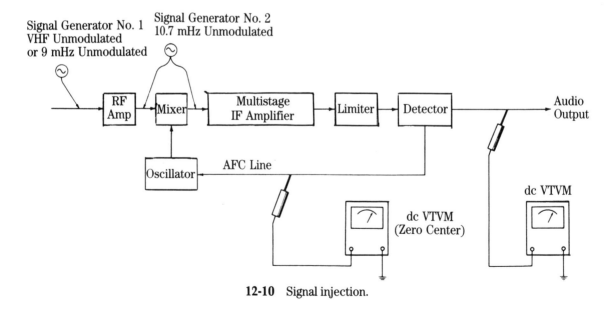

12-10   Signal injection.

grade instrument provided that it has reasonable short-term stability and has the capability of turning off the internal (usually AM) modulation. One problem associated with using low-cost signal generators is the poor dial accuracy that occurs at both 10.7 MHz and VHF. Also, most of these instruments use harmonics to provide the frequencies above 20 or 30 MHz. This can be a problem if those harmonics are of insufficient strength.

An alternative is to use a small transistorized crystal-controlled oscillator. The output buffer amplifier should be of sufficient power to give harmonics reasonable strength. Also, the output should be distorted to raise the harmonic content of the waveform. This approach is preferred by at least one car radio manufacturer.

The oscillator will give both good stability and close control of frequency accuracy. The 10.7 MHz crystal is, of course, for FM IF alignment. The 9 MHz crystal is used to provide FM band markers. Harmonics of 9 MHz will appear at 90, 99, and 108 MHz; this is perfect for most dial calibration purposes.

The RF signal should be injected into the system at the antenna jack via a suitable dummy antenna. The FM IF signal can be injected in one of three places, depending upon convenience and the mechanical layout of the radio. One of the best places is the base of the mixer transistor. Another point is the base of the input IF tube, transistor, or integrated circuit.

Another alternative, if either of the two preferred points prove inaccessible, is the same type of "gimmick" illustrated previously in Fig. 12-3C. That is, use a 2- or 3-inch piece of insulated hookup wire to "spray" the signal into the first FM IF transformer.

Two VTVM connections are needed. One is a zero-center VTVM across the automatic frequency control line. The other is a normal VTVM across a point to ground where the dc level varies with signal strength. In some sets this will be in the detector, while in others it will be either the base or the emitter of the limiter transistor. Most bench VTVMs can be made zero-center simply by adjustment of the "zero" or "left marker" controls.

Figure 12-11 shows a typical FM ratio detector circuit. Various alignment points are marked. The voltage across the AM suppression capacitor (point C) will indicate the relative strength of the signal. The IF transformers are peaked, using the voltage at this point as an indicator. The single most important adjustment is the secondary of the detector transformer in both discriminator and ratio detector designs. This is generally done by monitoring the AFC control line voltage.

When the input signal is unmodulated, the voltage produced on this line will have a certain discrete level when no signal is present. This value in most sets is zero. When any frequency other than 10.7 MHz is present, or when the detector transformer secondary is misaligned, there will be either a positive or negative voltage generated. This generation depends on whether the error is above or below the design frequency.

The secondary of the detector transformer is adjusted while monitoring the AFC line with a zero-center meter. At the same time, a 10.7 MHz continuous signal is applied to the input of the IF amplifier strip. Adjust the core of the transformer secondary for zero voltage along this line.

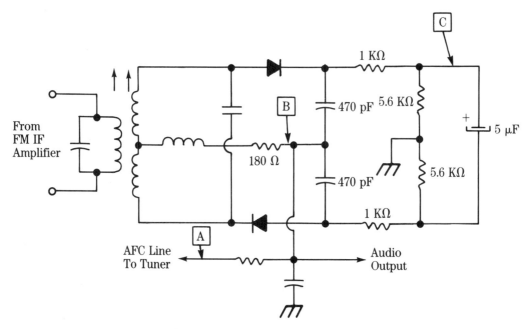

12-11 FM radio detector.

The IF transformers are aligned by adjusting their slugs while monitoring the signal level at an appropriate point. Peak the transformers for maximum signal level at the output. Again, the signal generator should be unmodulated.

Use the 9 MHz crystal to align the FM local oscillator. Set the tuner to either the 90, 99, or 108 MHz dial calibration points. Monitor the AFC control voltage with a zero-center meter. Turn either the local oscillator coil adjustment or trimmer-capacitor adjustment, depending upon the recommendation of the set maker, until the signal from the crystal oscillator or signal generator is centered in the passband of the radio. This will be indicated by the "zero" reading on the voltmeter.

It may be necessary to begin the alignment procedure on a higher voltmeter scale, repeating the process on lower scales until no further accuracy can be realized. As the adjustment is turned back and forth across the proper point, the zero-center voltmeter will deflect first in one direction and then in the other. This indicates that the performance of the detector is normal.

The crystal oscillator output will probably be too high for accurate alignment of the RF amplifier trimmers. Most better quality VHF signal generators, however, have a variable attenuator capable of reducing the output sufficiently to perform these adjustments. In a pinch you can use a weak signal off the air for this step. Peak the trimmers for the maximum signal output.

Alignment using a sweep generator

One good method for equipment hookup during sweep alignment of an FM car radio is shown in Fig. 12-12. Besides the sweep generator you will need a marker generator. This signal source will ideally produce both a 10.7 MHz and 100 KHz

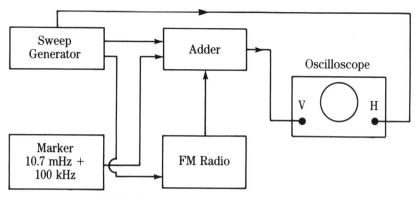

12-12 Sweep alignment setup for FM radio.

signal. The 10.7 MHz signal is needed to identify the center of the IF passband. The 100 KHz markers help locate subsidiary points within that passband.

The three signals, 10.7 MHz, 100 KHz, and the swept VHF FM signal, are combined in a piece of equipment called an adder. This device prevents interaction between the signals that could ruin the alignment. Some adders allow the two markers to appear on the scope trace horizontally so that they can be more easily identified. The current trend among test equipment makers is to incorporate all three items—sweep, marker, and adder—in one piece of equipment.

Figure 12-13 shows the oscilloscope traces that will appear during FM alignment. Figure 12-13A is the well-known discriminator S-curve that shows the varying response of the detector to frequencies within the receiver passband. The secondary (and to some extent the primary) of the FM detector transformer is adjusted for best symmetry of the S-curve.

This curve is taken across the discriminator output. The 10.7 MHz marker should be centered exactly on the zero line. The location of the other markers tends to vary somewhat from manufacturer to manufacturer. The frequencies illustrated, however, are a reasonably common recommendation.

The passband curve shown in Fig. 12-13B is taken from the IF amplifier. Ideally there will be no dip in the top portion of the curve. In reality however, there will be such a dip. This dip should be minimized during alignment so that it represents no

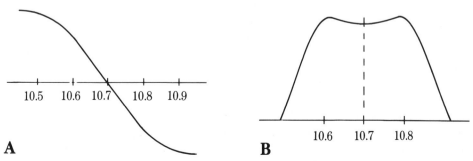

12-13 (A) Discriminator frequency response curve. (B) IF frequency response curve.

more than 10 percent of the total height of the curve as it appears on the scope screen.

Another ideal rarely achieved is best symmetry and maximum height, both occurring at the same adjustment points of the various IF transformers. Where there has to be a trade-off between symmetry of the curve and overall height, make it a rule of thumb to choose symmetry over gain. This advice will hold true except in areas that are a great distance from the nearest FM transmitter.

Best symmetry of the curve will occur when the −6 dB frequencies are in the correct place. These frequencies may differ from one brand or model to another. Typical instructions for monaural are 150 to 200 KHz. A stereo FM receiver should have a wider bandpass to accommodate the higher frequency sideband products generated by the encoded stereo information. On these, a passband with −6 dB points of 240 to 260 are generally acceptable.

Stereo Section Alignment

Although it is frequently performed incorrectly, the alignment of the multiplex decoder is as simple as aligning an AM radio. The principal tools are a stereo generator and an oscilloscope.

The stereo generator should produce the needed 19 KHz, 38 KHz, and 67 KHz signals; a stereo composite; and a VHF FM signal modulated by both the composite and the 19 KHz signals. The oscilloscope can be almost any audio type that will allow viewing signals up to 67 KHz. Just about every scope found in service shops will be more than sufficient for multiplex alignment.

The composite output of the multiplex generator can be injected directly into the stereo decoder input section. It would be better, however, to use it to modulate an FM generator, built into most multiplex generators, that feeds a signal to the antenna input. Otherwise the RC time constant of the decoder section input circuitry might change the composite signal enough to upset the alignment.

Adjust the SCA traps first. Modulate the generator with the 67 KHz audio signal. Place either the scope probe or the probe from an ac VTVM at point A in Fig. 12-14. Adjust the SCA trap so that a minimum (ideally no) signal appears at the test point.

Next, switch the generator so that it produces a 19 KHz output. Place the scope probe at any of the points indicated in Fig. 12-11. Peak the 19 KHz and 39 KHz transformers for maximum signal. If the scope probe is placed at a point early in the chain, it must be moved further toward the decoder circuit with each subsequent adjustment. Repeat the process until an increase in signal levels cannot be obtained.

The secondary of the 38 KHz output transformer should be adjusted with a stereo generator producing either a right only or left only signal. The phase of the 38 KHz signal, as applied to the decoder matrix, has a profound effect on the degree of channel separation produced. Adjust the output transformer secondary so that a maximum signal is produced in the active channel and the signal in the silent channel is minimized. At this time, any separation controls being used should also be adjusted in the same manner.

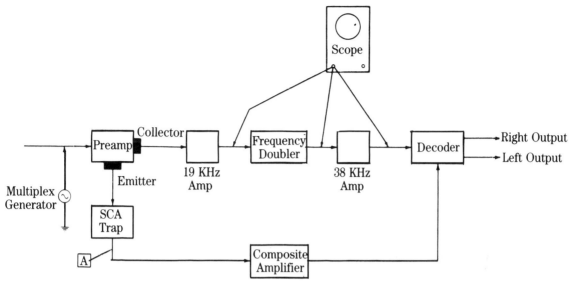

12-14 Oscilloscope testing of FM receiver.

Certain older stereo designs used a phase-locked oscillator to generate the 38 KHz decoding signal. In these sets use oscilloscope Lissajous patterns to adjust the oscillator frequency (Fig. 12-15).

Connect a sample of the 39 KHz signal to the vertical input of the oscilloscope. Connect the audio output of the stereo generator or the output of an accurate audio generator to the horizontal input of the scope. Set the external oscillator either to 38 KHz or to a frequency that is either a harmonic or subharmonic of 38 KHz. Suitable frequencies are 19 KHz, 38 KHz, or 76 KHz. Adjust the oscillator coil in the multiplex decoder until the Lissajous pattern on the screen of the oscilloscope locks in and remains stable.

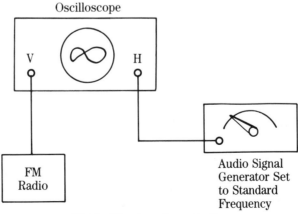

12-15 Lissajous figure testing.

Tune the radio on and off several stereo stations (especially a weak station) to be sure that the oscillator will lock properly every time. If it does not, try adjusting the locked-oscillator coil until the real stable point is found.

Alignment Tools

Radio alignment adjustments are not carried out with ordinary screwdrivers or similar tools. The metallic content of those tools upsets the tuned circuits being aligned, resulting in worse overall alignment than before the attempt was made. Special nonmetallic alignment tools, popularly called *diddlesticks*, are used.

Figure 12-16 shows several of the many available types. These tools are often available in multitool packs at low cost. There is one rule that most experienced radio technicians will agree on: it is hard to have too many varieties of diddlesticks! Several types of diddlestick are "must have," while others are highly desirable.

First, get a dual hexagonal diddlestick (Fig. 12-16A) in each of the three popular sizes. The hex tool is used to align those slug-tuned coil cores that have a female hex slug. The hex diddlestick comes in several different combinations of three sizes. Because the larger sizes are obsolete, they are not always easily obtained—but are the very ones needed in alignment of antique and classic radio receivers.

Also obtain a blade diddlestick (Fig. 12-16B). These tools are all plastic except for a pair of small metallic blades at each end. One blade will be exposed, while the

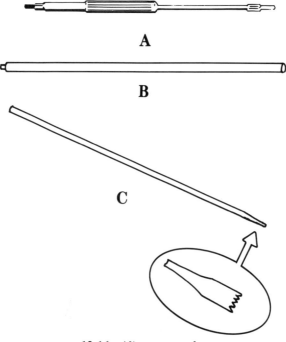

12-16 Alignment tools.

other is recessed. These tools are especially useful for adjusting trimmer capacitors and variable inductors that have a screwdriver slot cut into the end of a threaded brass rod attached to the slug-tuning core.

Finally, obtain a fiberglass diddlestick (Fig. 12-16C). These are long, thin fiberglass cylindrical rods that come in several sizes. Their purpose is to create special diddlestick shapes by filing or grinding a bevel or other shape on the end. A fine file or sandpaper is used to cut the end. These tools are used for slug-tuned coils that use a recessed rectangular slot in the slug for adjustment.

Figure 12-17 shows a useful little tool called the "magic wand." This tool is like an alignment tool, but it is tipped on one end with powdered iron, and on the other end with brass. When used as inductor cores these two materials have opposite effects on the inductance of the coil.

When a brass core is inserted into the coil form, the inductance will decrease—and the resonant frequency of the LC combination increases. When the iron core is inserted into the form the inductance increases—the resonant frequency of the LC combination decreases. The magic wand is used by the technician for finding out the direction a coil has to be adjusted in order to find resonance.

Conclusion

The alignment of radio receivers is a task that must be done from time to time. Far from being an arcane art, it is a learnable skill that should be added to your own "armamentarium" of radio servicing skills.

Brass Tip

Powdered Iron Tip

12-17 "Magic wand" tester.

13
Repairing Water
Damage to Radios

FIVE FULL DAYS OF HEAVY RAINS PELTED THE EAST COAST. IN OUR AREA THE RIVER crested 11 feet above flood stage, but eighty miles upstream in the narrow mountain canyons it became a 54-foot-high wall of water that overwhelmed the best efforts of hundreds of bone-tired volunteers. Despite heroic efforts, the sandbag wall at the edge of one town gave way under the relentless pressure of an angry river.

Over the next 24 hours the water rose, completely flooded basements and gushed into the first floor of most homes and businesses up to a height of six feet. As the waters receded, the governor called out the National Guard to prevent looters, and people returned home to recover what they could.

After cleaning out the water moccasins that came with flood waters, they found their possessions soaked and mudcaked. Among the damaged goods were radios, which they brought to the servicer in hopes that something could be salvaged. Could you help such a person?

Or could you even help yourself? Valuable radio collections can be lost completely. Service businesses can be forced into bankruptcy if their crucial test equipment and instruments are flood damaged—and the insurance is not enough. Economic survival might depend upon knowing how to handle flood damaged electronic equipment!

Although most flood damage scenarios are not as dramatic as this, we nonetheless hear of electronic equipment that has taken a bath: boating accidents, plumbing failures, and a variety of other problems splash equipment out of service.

I recall one incidence where a hospital plumber burst a 3.5-inch water pipe that he was repairing in a nursing station work room. Water came pouring out of the pipe at a high rate, causing a massive flood that damaged patient monitoring equipment in the OR and Post Anesthesia Recovery Room on the floor below.

After the smirks died over the announcement, "All housekeepers STAT to 2 East," a major effort was undertaken to save nearly $50,000 worth of electronic equipment that was freshwater damaged. Fortunately, there are certain things that a skilled technician can do to restore operation.

If the insurance company pays off well enough, then one could go out and buy a new product. But if the insurance company refuses to pay ("Sorry—wind-driven water damage excluded . . ."), or if there is no insurance, then you might want to attempt restorative action.

Even if the insurance company does pay off, customers can often buy back the equipment from them for salvage value. I recall one customer, a doctor, who received $325 for a 2-year-old marine VHF-FM two-way radio and bought it back from the insurance company for $20. The company sent him a check for $305, and he kept (and paid to restore) the "carcass."

Some of the steps recommended may sound a little bizarre but can restore an expensive piece of equipment. Some steps might cause a little damage that will also have to be repaired, especially those involving baking moisture out or using chemicals for cleaning. If that makes you nervous, then just remember that, you cannot harm the equipment anymore: it is already a total loss! Any action leading to restoration can result in only gain.

Before making any promises, however, make sure that the customer understands that you are undertaking "heroic" measures that may not be successful. One of the most frequently cited causes of bitter customer dissatisfaction is not your poor performance, but rather dashed expectations. If your customer is led to believe that the job will turn out much better than is possible, then he or she will not be in a forgiving mood when you fail to catch the bullet in your teeth.

That fact is especially true in medical service where egos are large and memories long. But if the job is a lot better than their expectations, then you will probably hear word of mouth "advertising" around town about your ability to walk on (or at least get rid of) water.

The first thing to do is refrain from turning the equipment on, even for a brief test to see if it is broken. Satisfy yourself right now that even a short dunk will cause fatal damage! Still the all-too-natural urge is to see if the equipment survived the flood: if it was immersed in water, then it did not survive.

Cleaning Steps

The first job is to remove the covers and give the equipment a bath. A shop in a seaport town finds saltwater damage to electronic equipment a common problem. One such shop received an $1800 UHF-FM radiotelephone set, the kind typically used in taxicabs, police cars, fire trucks, and ambulances, that had been immersed the night before during a storm.

The first thing the technician did was take the transceiver out on the back parking lot and give it a ten minute shower with a garden hose. He had lived in that town all his life, and therefore had much experience with water-damaged radio gear. If the damage is due to saltwater, then do the cleaning job immediately. The longer salt residue remains in the equipment, the greater will be the corrosion damage, and the lower the chance of successful restoration.

In some cases, it will be necessary to follow the shower with an immersion bath. One technician I know uses a 25-gallon tub, the kind you might use to give a large dog a bath. He mixes two to four quarts of a product like Lestoil, a small bottle (2 to 4 fl. oz.) of either fingernail polish remover or acetone (same chemical) and enough tap water to fill the tub all the way to the rim.

Leave the set in this bath for an hour, and then pour out the solution. Rinse the tub out thoroughly and refill with plain tap water. Some people prefer distilled water, which is available in bottles in some areas. This second bath removes the residue left by the chemicals in the first bath.

Note: This bath may damage some plastics. If this worries you, then use either plain soapy water or disassemble the plastic pieces and clean them by hand. It isn't quite as effective, but it works somewhat. Keep in mind that most plastic pieces can be replaced, and the damage will not usually prevent the set from operating: since it is already a total loss, don't worry about secondary damage!

Drying Procedures

The next step is drying the unit out thoroughly. If you live in Arizona, then simply leave the equipment out in the sun for about a week; everyone else will have to use some other method. The kitchen oven is a good bet, provided that it can be regulated to maintain a temperature of 125° to 130°F. That range is low for a kitchen oven, and some might not be able to remain that cool.

Higher temperatures will dry the set out faster, but they will also melt some of the plastics used in it, so beware. The drying process takes several days, perhaps as long as a week.

Another alternative is to build a box of cardboard, or other similar material, and use incandescent lamps of several hundred watts to provide heat. Use a thermometer inside the enclosure to ensure that the 130°F "melt limit" is not exceeded, and the box doesn't catch fire. Again, up to about a week is needed; although I recall one case in which a car radio dropped in a fresh water lake for a few minutes dried out in only one day.

Testing

Now comes the big test! In some cases, the only way to test the equipment is to turn it on and look for smoke. The more conservative approach sneaks up on it one step at a time. The first step in the test is to disconnect the dc power supply. This step can be absolutely essential to the future health of the set being repaired—especially those with high-voltage power supplies.

Without connecting the set to ac power, connect a bench power supply to the circuitry that was previously connected to the rig's internal power supply. It is essential that you use a dc power supply that will provide the same voltage(s) as the original internal supply, and which additionally (this is very important) has a current limiter control.

The output voltage is set to the dc voltage normally supplied by the set's power supply, and the current limiter control is set for a short-circuit current only a little above the normal operating current of the circuit under test.

Why go to such trouble? The reason is prevention of secondary damage. There is almost inevitably a short circuit or other condition that draws a large current. If such a condition exists in the equipment, then the normal internal power supply probably produces enough current to burn up components, printed wiring board tracks, and other components. After the circuit is checked out, then check out the power supply and, if working, reconnect it.

The low-voltage dc power supply should be checked out separately, especially if it uses a series-pass regulator, as most equipment does these days. If the voltage regulator circuit is not working, then several possible faults allow the rectifier output to be connected to the regulator output; this occurs when the series-pass transistor is either shorted or hard-biased fully on. Since the rectifier voltage is always higher than the regulator output voltage, it can damage circuits that were just pronounced healthy.

High-voltage power supplies have special problems all their own. These supplies are common in old radio sets. Small amounts of moisture that are no problem in low voltage supplies will permanently damage a high-voltage supply. The special problem is the high-voltage transformer. If moisture has penetrated the transformer, then the unit may have to be replaced. It may help to provide some extra drying for the transformer, but be prepared to replace it.

Figure 13-1 shows a method for drying a power transformer. A 115-volt ac lamp is placed in series with the primary of the high-voltage transformer. The current flow

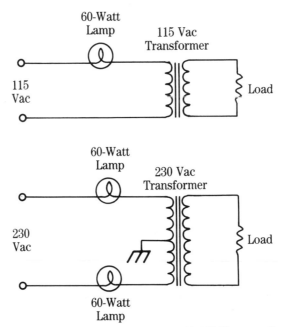

13-1 Circuit for drying out the radio transformer: (A) 115 Vac transformers, (B) 220 Vac transformers.

is enough to cause internal heat buildup, but not enough to destroy the transformer if it is shorted. If the high-voltage power supply uses a 220 Vac primary circuit, then place one lamp in series with each ac hot line (see Fig. 13-1).

Some remaining areas of concern, and probable damage, are those components whose moisture can get in and remain hidden. Candidates include: trimmer capacitors, air variable-capacitors, IF and RF transformers, switches and potentiometers, and paper and electrolytic capacitors.

With regard to trimmer capacitors, we can open the capacitor up to the minimum capacity position (screw all the way out) and apply a hair dryer or incandescent lamp for 10 or 15 minutes. Whether or not this step is needed can be determined after the initial power-on test shows a specific problem. Otherwise, you will destroy the alignment of the set for nothing. This step should not, therefore, be used merely as a matter of course—only in response to a specific symptom.

Likewise, air-variable capacitors may have corroded contact wipers between the rotor and stator, and this will become apparent when the rig is turned on.

Paper and electrolytic capacitors can absorb water, especially if they have a fiber or cardboard end-cap. If the capacitor shows signs of being soggy, then replace it; capacitors are, after all, relatively low-cost items.

If there remains a lot of scum on the printed wiring board, then spray clean it with Freon TF or some similar solvent. Some technicians prefer to use a small paint brush or "cheese cloth" to help remove the material.

Flood damaged equipment is often salvageable. The methods described above have been used by professional service technicians for many years—and are proved successful.

Recently the author heard from a reader who added some advice of his own. He was a former naval officer who used to have electronics technicians working for him on board a naval ship. He said they used to repair salt-water soaked equipment in an unusual manner. A sailor would take the equipment into the shower, and slosh it down with warm water.

They then took the desalinated equipment to the galley (kitchen) and dried it out in the ovens with low heat and good air circulation. The retired officer also advised that distilled water is best, and that tap water in some locations is too hard (i.e., contains minerals). Anyone using this method must either buy distilled water or use an in-line water softener.

For a chassis covered with oily dirt, the equipment can be cleaned with a mixture of 8 to 10 ounces of household ammonia, 4 to 6 ounces of a cleaner such as Mr. Clean or Lysol, 4 to 6 ounces of acetone (used in some fingernail polish removers), and enough distilled, or soft, water to make one gallon of solution. The equipment is dunked into this solution.

For larger equipment, proportionally larger amounts can be used. An old dental Water Pik can be used to hose off equipment that is too large to dunk. The equipment is then dried in an oven set to 140° to 150°F for 4 to 5 hours. (Note: some plastics used in electronic equipment will melt at temperatures over 130 degrees, so beware.) All lubricants in switches, potentiometers, and air variable-capacitors (where used) must be replaced after this treatment.

The black asphaltlike paste that oozes out of overheated transformers can be easily removed from chassis by using either freeze spray, or a blast from a CO_2 fire extinguisher. Use an underpressure one that already needs refilling; don't waste your protection on cleaning jobs. The frozen paste becomes brittle and can be flaked off using a dental tool or soldering aid tool.

Warning: mixing other cleaners with Chlorox® or other hypochlorite-containing liquids can release poisonous chlorine gas. Only do this job in a well-ventilated area, or outdoors.

14
Electrical Safety
on the Workbench

ELECTRICAL AND ELECTRONIC EQUIPMENT IS INHERENTLY DANGEROUS, SO IF IT IS not used according to some basic commonsense rules it can lead to injury or even deaths. The medical environment has some special problems regarding electrical safety because the normally protective skin of the patient is breeched. There are three situations to consider: burns, macroshock, and microshock.

Electrical Shock

Electrical shock incidents can cause burns of all three degrees. There are two basic mechanisms of burning. One is the flash that occurs when an electrical arc occurs. The other is when current flows through tissue.

Macroshock   Macroshock is the usual type of electrical shock that comes from direct contact with an electrical source. If you touch the 110 volt ac line while grounded, then a very painful and possibly fatal shock will occur. This is macroshock, and does not require wounds or other breeches of the skin.

Microshock   Microshock is a more subtle form of electrical shock, which not long ago was not recognized. But increased use of electrical equipment in hospitals in the 1950s and 1960s led some authorities to speculate that as many as 1200 people a year were being accidentally electrocuted in hospitals from tiny currents that went unnoticed by the medical staff. Microshock is electrical shock from minute currents that are too small to affect persons with intact skin, but will affect persons if they are introduced inside the body through a wound.

Other hazards   In all forms of electrical shock a difference in electrical potential must exist between two points on the body. In other words, two points of contact must exist between the victim and the electrical source. That is why you sometimes see harmless "hair raising" exhibits when a person touches an electro-

238

static high voltage and their hair stands on end. These potentials are monopolar with respect to the demonstrator, so no current flow exists.

Similarly, some less prudent electricians will work a circuit "hot," without turning off the power, in seeming safety because they take care to not ground themselves or in any other way come between the hot wire and ground, or across two hot wires. This is a bad practice that ought to be strongly discouraged.

In addition to electrical shock there are other safety concerns regarding electricity. One major problem is fire. Overloaded or defective electrical circuits can spark and create a fire. Many residential, business, industrial, and health care facility fires every year are traced to faulty wiring or malfunctioning electrical equipment.

Electrical faults will also damage the equipment, the building, or other equipment. A short circuit that is not protected by a fuse may create more damage to the shorted equipment, and may also cause damage to the building wiring and electrical components. In extreme cases, a fire may result. When fuses and circuit breakers are either not used, or defeated ("penny in the fuse box") a severe fire hazard exists, and any damage to equipment is certainly increased.

Less recognized, but nonetheless possible, is the hazard of explosion from electrical faults. An overloaded circuit or electrical component builds up internal pressure (often from gas released when the device is severely overheated) and ruptures spectacularly. Examples include high-power transformers and main capacitors inside large electronic devices.

The second explosion hazard is sparking in the presence of flammable gases or vapors. If an electrical circuit is disconnected while operating, or if certain faults exist, then a spark may result. If that spark occurs when either flammable gases, oxygen (which isn't flammable itself, but acts that way because it violently promotes burning of other materials), or vapors, such as gasoline or certain waxes, then a violent and dangerous explosion may result.

Besides the obvious danger of fragments from the casing of an exploding device, there is also the possibility of boiling oil splattering nearby personnel. There is a specific danger regarding this oil in some cases. Certain older capacitors and transformers were built using PCB oil as an internal coolant. PCB oil is a severe carcinogen. The importance of that statement cannot be underestimated: PCB is dangerous! Although most electrical devices using PCB oil are now out of service, it is possible that some are still around, especially in older equipment or buildings.

Equipment to be especially suspicious of includes elderly high-power RF generators. If one of these devices is found, then it is a good idea to have it checked by a component technician or engineer to determine if the large, high-voltage capacitor(s) inside uses PCB oil as an insulating coolant or dielectric material. A PCB spill can close a building for a long period of time until a proper cleanup is completed.

What To Do For An Electrical Shock Victim

Before discussing either form of electrical shock, let's discuss the mechanism of electrical shock and the proper approach to dealing with a shock victim.

The usual cause of death from electrical shock is a phenomenon called *ventricular fibrillation*. ("V. Fib."). This is an arrhythmic heartbeat in which the heart merely quivers instead of beating. V.Fib. is incapable of sustaining blood pumping effectiveness, so the victim dies within a few minutes unless someone trained in cardiopulmonary resuscitation (CPR) is nearby.

Before you can aid the victim of electrical shock you must be sure that either the victim is away from the current, or the current is turned off. Otherwise, when you touch the victim, you will also become a victim!

As soon as the victim is clear of the electrical current, initiate cardiopulmonary resuscitation (CPR), and send for help. CPR will not bring the victim out of V.Fib. Its function is to provide life support until properly equipped and trained medical personnel can be summoned. They will use a defibrillator to shock the victim's heart back into correct rhythm along with drugs and intravenous solutions in order to re-establish the body's balance.

None of these actions can be performed by the untrained person. Even though CPR cannot be effectively performed by the untrained person, everyone who works near, on, or around electrical equipment should learn it. In addition, teenage and adult family members should also learn CPR; after all, who is going to save you if the electrical accident occurs at home?

The local Red Cross, the Heart Association, some community colleges, and most local hospitals can direct you to certified CPR courses. It is impossible for you to learn CPR from watching medical shows on TV, so get trained by a knowledgeable instructor.

How Much Current Is Fatal?

One day I overheard a hospital interm claim that 110 volts ac from the wall socket is not dangerous because they told him in medical school that it's not the voltage that kills, it's the current.

I asked him: "Doctor, have you ever heard of Ohm's law?" According to Ohm's law, $I = E/R$.

A statistic that the doctor apparently didn't know was that 110 volts ac from residential wall sockets is the most common cause of electrocution in the United States. In addition, medical studies reveal that the 50 to 60 Hz frequency used in ac power distribution almost worldwide is the most dangerous range of frequencies.

Higher and lower ac frequencies are less dangerous than 60 Hz ac (but not safe!) According to medical experts who have studied electrical shock, the killing factor is *current density* in a certain area in the right atrium of the heart called the *sinoatrial node*.

Current flowing through that section of the heart can induce fatal V. Fib. In general, for limb-contact electrical shocks through intact skin (macroshock), the following approximations are accepted:

1–5 mA	Level of perception
10 mA	Level of pain
100 mA	Severe muscular contraction
100–300 mA	Electrocution

Keep in mind that these figures are approximate, and not to be taken as guidelines to approximate "assumed risk." Death can occur under certain circumstances with considerably lower levels of current. For example, when you are sweating, or are standing in salt water, then risks escalate tremendously.

In medical situations, the level of current that can kill us in the 20–150-microampere level because the current is induced directly into the body, which is essentially a "salt water" environment. Human skin has a resistance of 1000 to 20,000 ohms, while internal tissue has a resistance of around 50 ohms. Microshock is discussed below.

Is High Current, Low Voltage Safe?

I once attended a design review meeting for a high-power mobile transmitter, such as the type used in ambulance and police cars. The engineer's specifications called for installation of low-voltage (28 Vdc), high-current (30-ampere) dc power-supply terminal insulation. One of the engineers present sneered that that was like asking him to insulate the battery terminals of his car. He implied that low voltage can never hurt you. There are two false premises to that opinion.

First, although low-voltage, high-current points rarely cause electrical shock, it is possible for a dangerous shock to occur when the person has a very low skin resistance (is very sweaty) or has an open wound. Although the case did not result in electrocution, an electronics technician I know recently injured himself severely when he cut himself on a +5 Vdc, 30-ampere computer power-supply terminal. A large amount of current flowed in his arm, and caused severe pain and some physical damage.

Second, high current is extremely dangerous if you happen to be wearing jewelry. A two-way radio shop used 12-volt batteries and battery chargers for the troubleshooting bench supply for mobile service. A technician working on the battery rack dropped a wrench onto the battery, making contact from (−) to (+) through his watchband. The large current turned the watchband red hot, and gave him one serious second and third degree burn.

Don't assume that low-voltage high-current power supplies are harmless!

Common Hazardous Electrical Shock Situations

In order to raise our consciousness about how shock can occur let's take a look at certain scenarios of electrical shock that might occur with electronics service technicians. Figure 14-1 shows the direct approach to fatal electrical shock. You are grounded through conductive shoes and touch an electrically hot point. You need not be outdoors; a concrete garage, shop, or basement floor is a reasonably good conductor, as are wet leather and some rubber shoes.

Figure 14-2 shows an indirect scenario that especially affects electronics workers. Consider the grounded instrument probe—in this case an oscilloscope. When you grasp that probe, you may be grounded through the scope shield and the

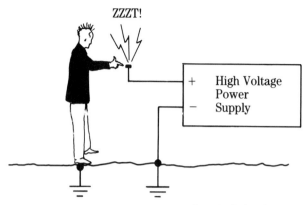

14-1 Scenario guaranteeing electrical shock.

power cord ground conductor. If you then touch a "hot" point, you will get shocked, and may be killed.

A related scenario is shown in Fig. 14-3. Here we have an ac/dc consumer appliance, such as a low-cost radio or TV set. Note that the oscilloscope probe ground is connected to the set ground, which also happens to be one side of the ac power line.

Everything is fine as long as the ac plug is oriented correctly in the wall, and if the wall socket is wired correctly. But if you plug it into the wall receptacle backwards, then there will be an explosive short circuit and possible electrocution of the operator.

Another scenario is the fatal antenna erection job. It is NEVER good practice to erect an antenna near a power line. Every year we hear stories of people electrocuted because either an antenna they were working on fell across the power lines, or they tried to toss a wire antenna over the power line in order to raise the antenna above the lines, or a ladder they were using fell across the power lines.

These tactics will kill you. Incidentally, this is the reason why OSHA-approved industrial ladders are made of wood or other nonconductive material, not of aluminum as consumer ladders.

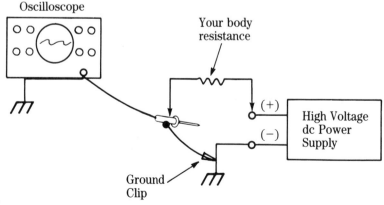

14-2 Hidden scenario that is just as dangerous!

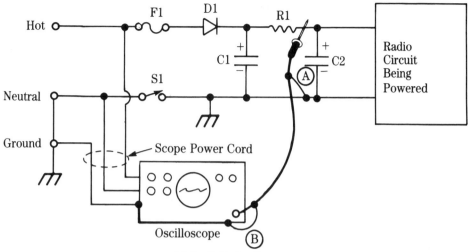

14-3 Path for disaster.

Some Cures for the Problems

Figure 14-4 shows the schematic for the usual U.S. residential ac electrical system. Industrial electrical systems as used in hospitals are a bit different at the service entrance, but become much like Fig. 14-4 when the power is distributed throughout the building.

The power company distributes energy through high-voltage lines. When it arrives at a point a short distance from the customer, it is stepped down in a "pole pig" transformer to 220 Vac center-tapped. The center-tap of the transformer secondary is grounded, and therein lies the problem.

The two ends of the 220 Vac secondary are brought into the building as a pair of 110 Vac hot lines. Tapping across the two lines produces a 220 Vac outlet; tapping from the ground line—the transformer C.T.—to either hot line produces a 110 Vac outlet.

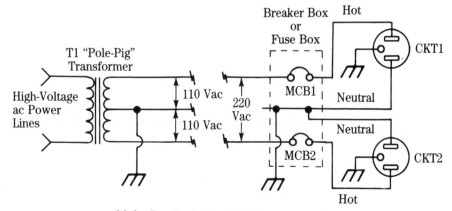

14-4 Standard ac residential power system.

The problem is that the electrical system in the United States is ground referenced. The solution is to make the local electrical system nonground referenced. This method is used in hospital operating rooms and in some intensive care units for patient safety reasons. It should also be used on radio service benches, especially if ac/dc sets are serviced.

Figure 14-5A shows the wiring for such a system. Transformer T1 is one of two forms of isolation transformer: a 1:1 transformer gives a 110 Vac isolated (nonground referenced) ac line from a 110 Vac standard line; a 2:1 transformer does the same thing from a 220 Vac line.

The second transformer, T2, is an *autotransformer*, used for varying the voltage on the ac line. It will typically allow you to set the output voltage from 90 Vac to 140 Vac with a 110 Vac line-voltage applied. This transformer is used by servicers to set

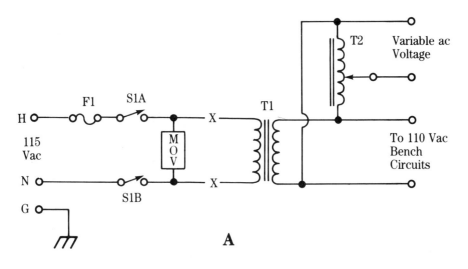

A

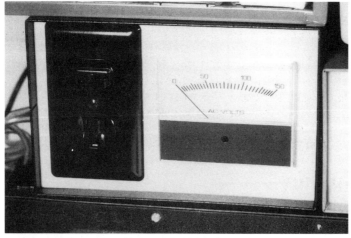

14-5 (A) Variable voltage isolated ac power system for the bench. (B) Isolation transformer for the bench.

the voltage higher or lower than normal to check radio operation or expose problems.

If you work on either radio transmitters or other high-power RF-producing devices, such as electrosurgery machines, or work near such generators, then you might want to place an electromagnetic interference (EMI) filter in the line at the points marked X. The EMI filter is an LC-section that attenuates RF, but not 60 Hz power.

The MOV, or metal oxide varistor, is used to clip the amplitude of high-voltage line transients or around 100 microseconds, that could either damage or interfere with the operation of the equipment on the bench.

The circuit breaker or fuse is used to protect equipment on the bench, as well as the transformer. It is always placed in the hot line, or in both lines. Fuses and circuit breakers are never placed in the neutral line only. The switching shown in Fig. 14-5A breaks both lines. I prefer this approach on the theory that hot and neutral lines can be reversed accidentally, and leave you in the position of breaking a neutral, and leaving the hot line alive.

Figure 14-5B shows an isolation transformer intended for service bench use. I use this device on my workbench in order to ensure safety.

Some General Advice on Safety

There is only one way to ensure that the ac line won't shock you: disconnect it. Make it your practice to *never* work on equipment that has the plug inserted into the power outlet. Don't trust switches, fuses, circuit breakers, or other people.

You might have heard that you should work on high-voltage devices with your left hand in your pants pocket. That advice is based on the theory that the "lefthand to either leg path" is supposedly the most deadly. But placing one hand in your pocket leaves you awkward; you are unable to safely work on the circuit with only one hand. It is better to use both hands, so that the work environment is safe.

What is a safe work environment? The power system should be isolated (as discussed above). The floor should be insulated by a carpet, treated masonite, a plastic cover, a rubber mat, wooden planking, or some other material; the floor should always be well insulated and kept dry. An isolation transformer should be used on the workbench for servicing radios.

When working on high-voltage dc circuits, keep in mind that capacitors store electrical charge. All filter capacitors must be discharged manually after the power is turned off. Also, the capacitor must be discharged multiple times. Even when the capacitor terminals are shorted, not all the energy is removed the first time. Some energy is stored in the dielectric even after the main charge is removed.

Conclusion

Unfortunately, since electricity is invisible, there is no such thing as complete, failure-free safety with electrical devices. But proper recognition of the mechanisms of danger and proper management of the risks will ensure that the environment is as safe as possible.

Appendix
Component color code

B   For other electrical and electronics symbols refer to military standard, MIL-STD-15-IA

C

D   Standard colors used in chassis wiring for the purpose of circuit identification of the equipment are as follows:

E

Color	1st digit	2nd digit	Multiplier	Tolerance (percent)
Black	0	0	1	
Brown	1	1	10	
Red	2	2	100	
Orange	3	3	1,000	
Yellow	4	4	10,000	
Green	5	5	100,000	
Blue	6	6	1,000,000	
Violet	7	7	10,000,000	
White	9	9	1,000,000,000	
Gold			.1	5
Silver			.01	10
No color				20

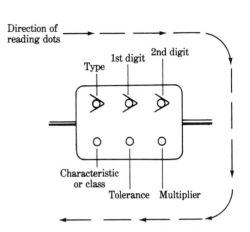

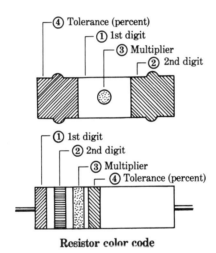

Resistor color code

Type	Color	1st digit	2nd digit	Multiplier	Tolerance (percent)	Characteristic or class
JAN, MICA	Black	0	0	1.0	± 1	Applies to temperature coefficient or methods of testing
	Brown	1	1	10	± 2	
	Red	2	2	100	± 3	
	Orange	3	3	1,000	± 4	
	Yellow	4	4	10,000	± 5	
	Green	5	5	100,000	± 6	
	Blue	6	6	1,000,000	± 7	
	Violet	7	7	10,000,000	± 8	
	Gray	8	8	100,000,000	± 9	
EIA, MICA	White	9	9	1,000,000,000		
	Gold			.1	±10	
Molded paper	Silver			.01	±20	
	Body					

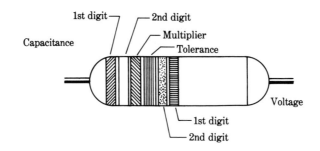

	Capacitance			Tolerance (percent)	Voltage rating	
Color	1st digit	2nd digit	Multiplier		1st digit	2nd digit
Black	0	0	1	±20	0	0
Brown	1	1	10		1	1
Red	2	2	100		2	2
Orange	3	3	1,000	±30	3	3
Yellow	4	4	10,000	±40	4	4
Green	5	5	100,000	± 5	5	5
Blue	6	6	1,000,000		6	6
Violet	7	7			7	7
Gray	8	8			8	8
White	9	9		±10	9	9

6-Band color code for tubular paper dielectric capacitors

B - A - Temperature coefficient
B - 1st digit
C - 2nd digit
D - Multiplier
E - Tolerance

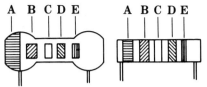

Radial Lead Ceramics

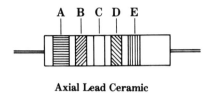

Axial Lead Ceramic

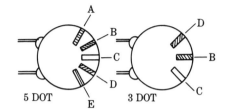

5 DOT   3 DOT

Ceramic Disc Capacitor Marking

Color	1st digit	2nd digit	Multiplier	Tolerance More than 10 pf (in percent)	Tolerance Less than 10 pf (in pf)	Temperature coefficient*
Black	0	0	1.0	±20	±2.0	0
Brown	1	1	10	± 1		−30
Red	2	2	100	± 2		−80
Orange	3	3	1,000			−150
Yellow	4	4	10,000			−220
Green	5	5		± 5	±0.5	−330
Blue	6	6				−470
Violet	7	7				−750
Gray	8	8	.01		±0.25	+30
White	9	9	.1	±10	±1.0	+120 TO −750 (EIA) +500 TO −330 (JAN)
Silver						+100 (JAN)
Gold						Bypass or coupling (EIA)

*Parts per million per degree centigrade.

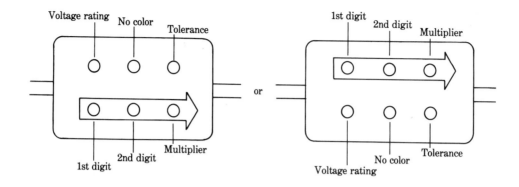

Color	1st digit	2nd digit	Multiplier	Tolerance (percent)	Voltage rating
Black	0	0	1.0		
Brown	1	1	10	± 1	100
Red	2	2	100	± 2	200
Orange	3	3	1,000	± 3	300
Yellow	4	4	10,000	∓ 4	400
Green	5	5	100,000	± 5	500
Blue	6	6	1,000,000	± 6	600
Violet	7	7	10,000,000	± 7	700
Gray	8	8	100,000,000	± 8	800
White	9	9	1,000,000,000	± 9	900
Gold			.1		1000
Silver			.01	±10	2000
Body				±20	*

*Where no color is indicated, the voltage rating may be as low as 300 volts.

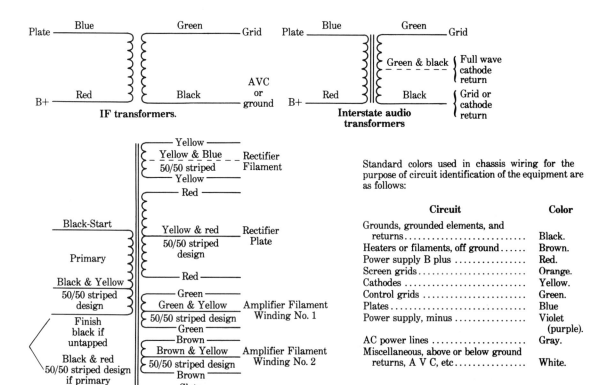

IF transformers.

Interstate audio transformers

Power transformers

Standard colors used in chassis wiring for the purpose of circuit identification of the equipment are as follows:

Circuit	Color
Grounds, grounded elements, and returns	Black.
Heaters or filaments, off ground	Brown.
Power supply B plus	Red.
Screen grids	Orange.
Cathodes	Yellow.
Control grids	Green.
Plates	Blue
Power supply, minus	Violet (purple).
AC power lines	Gray.
Miscellaneous, above or below ground returns, A V C, etc	White.

For other electrical and electronic symbols refer to military standard, MIL-STD-15-IA

Index